EVERYDAY AUTOMATION

This Open Access book brings the experiences of automation as part of quotidian life into focus. It asks how, where and when automated technologies and systems are emerging in everyday life across different global regions? What are their likely impacts in the present and future? How do engineers, policy makers, industry stakeholders and designers envisage artificial intelligence (AI) and automated decision-making (ADM) as solutions to individual and societal problems? How do these future visions compare with the everyday realities, power relations and social inequalities in which AI and ADM are experienced? What do people know about automation and what are their experiences of engaging with 'actually existing' AI and ADM technologies? An international team of leading scholars brings together research developed across anthropology, sociology, media and communication studies and ethnology, which shows how by rehumanising automation, we can gain deeper understandings of its societal impacts.

Sarah Pink is Professor at Monash University Australia, where she is Director of the Emerging Technologies Research Lab, Associate Director of the Monash Energy Institute and an investigator in the Australian Research Council Centre of Excellence for Automated Decision-Making and Society.

Martin Berg is Professor of Media Technology at Malmö University, Sweden. He coordinates the Riksbankens Jubileumsfond's research network Re-humanising Automated Decision-Making and co-directs Malmö University's strategic research programme Data Society.

Deborah Lupton is SHARP Professor in the Centre of Social Research in Health and Social Policy Research Centre, Faculty of Arts, Design and Architecture, UNSW Sydney. She leads the Vitalities Lab and the UNSW Node of the Australian Research Council Centre of Excellence for Automated Decision-Making and Society.

Minna Ruckenstein is Associate Professor in the Centre for Consumer Society Research, University of Helsinki. She leads an interdisciplinary research group that studies algorithmic culture and organisational and societal processes in relation to automated decision-making.

EVERYDAY AUTOMATION

Experiencing and Anticipating Emerging Technologies

Edited by Sarah Pink, Martin Berg, Deborah Lupton and Minna Ruckenstein

Routledge
Taylor & Francis Group

LONDON AND NEW YORK

Cover image: Kwanchai Lerttanapunyaporn/EyeEm/Getty Images

First published 2022
by Routledge
4 Park Square, Milton Park, Abingdon, Oxon OX14 4RN

and by Routledge
605 Third Avenue, New York, NY 10158

Routledge is an imprint of the Taylor & Francis Group, an informa business

British Library Cataloguing-in-Publication Data
A catalogue record for this book is available from the British Library

Library of Congress Cataloging-in-Publication Data
Names: Pink, Sarah, editor.
Title: Everyday automation : experiencing and anticipating emerging
   technologies / edited by Sarah Pink, Martin Berg, Deborah Lupton,
   Minna Ruckenstein.
Description: Abingdon, Oxon ; New York, NY : Routledge, 2022. |
   Includes bibliographical references and index.
Identifiers: LCCN 2021056583 (print) | LCCN 2021056584 (ebook) |
   ISBN 9780367773403 (hardback) | ISBN 9780367773380 (paperback) |
   ISBN 9781003170884 (ebook)
Subjects: LCSH: Technology—Social aspects. | Automation—Social
   aspects. | Human-computer interaction-—Social aspects.
Classification: LCC T14.5 .E965 2022 (print) | LCC T14.5 (ebook) |
   DDC 303.48/3—dc23/eng/20220131
LC record available at https://lccn.loc.gov/2021056583
LC ebook record available at https://lccn.loc.gov/2021056584

ISBN: 978-0-367-77340-3 (hbk)
ISBN: 978-0-367-77338-0 (pbk)
ISBN: 978-1-003-17088-4 (ebk)

DOI: 10.4324/9781003170884

Typeset in Bembo
by Apex CoVantage, LLC

CONTENTS

FIGURES

CONTRIBUTORS

Mark Andrejevic is Professor in the School of Media, Film and Journalism, where he leads the Automated Society Working Group. He is a chief investigator and co-leader of the Data Program in the Australian Research Council Centre of Excellence for Automated Decision-Making and Society. His research focuses on digital media, surveillance and popular culture. He is the author, most recently, of *Automated Media* (2019).

Dan Angus is Professor of Digital Communication in the School of Communication and leader of the Computational Communication and Culture program in the Digital Media Research Centre, Queensland University of Technology. His research focuses on the development and application of computational analysis methods in communication and media studies, with a specific focus on conversation and social media data.

Martin Berg is Professor of Media Technology at Malmö University, Sweden. His research interests include digital sociology as well as critical studies of algorithms and automation processes. He currently leads the project 'Working with Algorithmic Colleagues: Expectations and Experiences of Automated Decision-Making' (funded by The Swedish Research Council from 2021 to 2024), coordinates the research network 'Re-humanising Automated Decision-Making' (funded by the Swedish foundation 'Riksbankens Jubileumsfond') and co-directs Malmö University's strategic research programme 'Data Society'. Berg's recent book *Imagining Personal Data: Experiences of Self-Tracking* (Routledge, 2020) was co-authored with Vaike Fors, Sarah Pink and Tom O'Dell.

Magnus Bergquist is Professor of Informatics at the School of IT, Halmstad University, Sweden. He has studied open source, open and distributed innovation, interorganisational communication, cultural aspects of IT, everyday IT use and the use of social media in the public sector. His recent research involves studies on digital service innovation for mobility and digitalisation in healthcare organisations. His research is published in journals such as *Information Systems Journal, Research Policy, Informatics for Health and Social Care, International Journal of Human–Computer Interaction* and *Computers and Education.*

Meike Brodersen is a postdoctoral researcher at the School of ITE at Halmstad University in Sweden. With a PhD in sociology from the Université Libre de Bruxelles, she has specialised in adapting ethnographic methods to pursue questions of urban studies and the sociology of mobility. She has a particular interest in automation and the implications of digitalisation in the context of mobile biographies and local working and mobility environments.

Robbie Fordyce is from Aotearoa and is Lecturer in Big Data/Quantitative Analytics and Research Methods at the School of Media, Film and Journalism at Monash University. He researches the exploits, manipulations and politics of rule-based systems and their cultures.

Vaike Fors is Professor of Design Ethnography at the School of ITE at Halmstad University in Sweden. Her area of expertise lies in the fields of visual, sensory and design ethnography. In her pursuit to contribute to further understandings of contemporary conditions for learning, she has studied people's interaction with new and emerging technologies in various research projects. She is an experienced project leader of international scientific, applied and collaborative research projects. Her recent publications include the co-authored books *Imagining Personal Data: Experiences of Self-Tracking* (Routledge, 2020) and *Design Ethnography* (2022).

Heather A. Horst is Professor and Director of the Institute for Culture and Society at Western Sydney University. A sociocultural anthropologist by training, she researches material culture and the mediation of social relations through digital media and technology. Her recent books include *The Moral Economy of Mobile Phones: Pacific Islands Perspectives, Location Technologies in International Context* and *Digital Media Practices in Households: Kinship Through Data.*

Anne Kaun is Professor in Media and Communication Studies at Södertörn University, Stockholm, Sweden. Her research areas include media theory, mediated temporalities, algorithmic culture, automation and artificial intelligence from a humanistic social science perspective. Her work has appeared in, among others, *New Media & Society; Media, Culture & Society;* and *Information, Communication & Society.* In 2020, her co-edited book *Making Time for Digital Lives* was published by Rowman & Littlefield.

Jenny Kennedy is Research Fellow in Media and Communication at RMIT University, Australia. She presently holds an Australian Research Council Discovery Early Career Research Award (ARC DECRA). She is Associate Investigator in the ARC Centre of Excellence for Automated Decision-Making and Society (ADM+S) and is Core Member of the Digital Ethnography Research Centre (DERC). Her research charts shifts in domestic digital technology practices against the context of rapid evolutions in digital infrastructures, smart devices, artificial intelligence and automated decision-making. Her most recent co-authored books include *Digital Domesticity: Media, Materiality, and Home Life* (Oxford University Press, 2020) and *The Smart Wife: Why Siri, Alexa, and Other Smart Home Devices Need a Feminist Reboot* (MIT Press) with Yolande Strengers.

Tuukka Lehtiniemi is a postdoctoral researcher at the Centre for Consumer Society Research, University of Helsinki. His research focuses on datafication, data activism and automated decision-making, with an aim to understand how our uses of digital technologies are shaped by how we imagine the economy to work. He holds a PhD degree in economic sociology.

Luzhou Li is Lecturer at Monash University and is Member of the Automated Society Working Group. Her research focuses on global media industries, political economy of media, media policy and governance and contemporary China. She is the author of *Zoning China: Online Video, Popular Culture, and the State* and currently holds an early career research fellowship from the Australian Research Council to study Chinese social media platforms.

Stine Lomborg is Associate Professor at the Department of Communication, University of Copenhagen. Her research focuses on digital tracking infrastructures, people's communication practices and engagement with tracking data, and what it means to live a good, safe and meaningful datafied life.

Deborah Lupton is SHARP Professor and Leader of the Vitalities Lab in the Centre for Social Research in Health and Social Policy Research Centre at UNSW Sydney, Australia. She is also a leader of the UNSW Node, Health Focus Area, and co-leader of the People Program in the Australian Research Council Centre of Excellence for Automated Decision-Making and Society. She is the author/co-author of 19 books and editor/co-editor of a further nine volumes, the latest of which are *The Face Mask in COVID Times* (de Gruyter, 2021), *The COVID-19 Crisis* (Routledge, 2021), *Creative Approaches to Health Education* (Routledge, 2022) and *COVID Societies* (Routledge, 2022).

Dick Magnusson is Associate Professor of Technology and Social Change at Linköping University, Sweden. His research focuses on sociotechnical system change, with specific interests in urban infrastructure, urban and regional planning aspects and grassroots innovations. He is the programme coordinator for the Bachelor Programme in Urban and Regional Planning at Linköping University.

Sheba Mohammid (PhD) is Institute Associate at the Institute for Culture and Society at Western Sydney University. She has over ten years of experience in national and global policy, engagement and research and was named an Emerging Leader of the Digital World (EC and Diplofoundation). She completed her PhD at RMIT University in 2017 on the topic of 'Digital media, learning and social confidence: An ethnography of a small island knowledge society'.

Sarah Pink is Professor and Director of the Emerging Technologies Research Lab at Monash University. She is also co-leader of the People Programme and Leader of the Transport Mobilities focus area of the Australian Research Council Centre of Excellence for Automated Decision-Making and Society and an associate director of the Monash Energy Institute. She is an interdisciplinary anthropologist blending ethnography, design and documentary in her theory and practice. Her recent book and film publications include the newly revised futures-focused *Doing Visual Ethnography* 4th edition (2021), the documentary film *Smart Homes for Seniors* (Pink Dir. 2021) and the co-authored *Design Ethnography* (2022).

Kaspar Raats is Industrial PhD Candidate at the School of ITE at Halmstad University and Volvo Cars in Sweden. His area of expertise lies within user experience research and design. His research focus is on experience of trust in intelligent vehicle technologies and services. Through experimenting with design ethnography, co-design and design fiction, he explores how to develop more transparent and human-centred algorithmic systems.

Lina Rahm is Doctor of Pedagogy and the Ragnar Holm research fellow at the Department of Philosophy and History of Technology at The Royal Institute of Technology, Stockholm, Sweden. Her research is focused on sociotechnical and educational imaginaries. Specifically, she has mapped out the genealogy of the digital citizen as an entanglement of automation, education and citizenship. Empirically, her research spans from the 1950s up until today.

Harald Rohracher is Professor of Technology and Social Change at Linköping University, Sweden. His research focuses on the governance of sociotechnical change towards greater sustainability, sustainable energy technologies, urban low-carbon transitions and the roles of users and civil society in innovation processes. He is an associate editor of *Environmental Innovation and Societal Transitions*.

Bertil Rolandsson is Associate Professor of Sociology at the Department of Sociology and Work Science, University of Gothenburg, Sweden. His research interests include digitalisation of work, AI and professionalism, the use of social media in the public sector and digital surveillance. He has published a range of studies in journals such as *Research Policy*, *Organization Studies*, *Tecnoscienza* and *Qualitative Research in Organizations and Management*. He currently coordinates an interdisciplinary network called 'Opportunities and challenges for the Nordic models in the digital era', funded by the Swedish Research Council for Health, Working Life and Welfare (FORTE), engaging 15 different researchers from all the Nordic countries.

Minna Ruckenstein is Associate Professor at the Centre for Consumer Society Research, University of Helsinki. She directs an interdisciplinary research group that explores economic, social, emotional and imaginary aspects of algorithmic systems and processes of datafication. Her ongoing projects focus on algorithmic culture and rehumanising automated decision-making.

Neil Selwyn is a distinguished research professor in Monash University's Faculty of Education. His research currently focuses on issues of datafication, digital labour and the AI-driven automation of education. His recent and forthcoming books include *Critical Digital Literacies* (MIT Press, 2023), *Should Robots Replace Teachers?* (2019, Polity) and *What Is Digital Sociology?* (2019, Polity). @neil_selwyn

Rachel Charlotte Smith is Associate Professor of Design Anthropology at the Research Centre for Participatory Information Technology (PIT), Aarhus University. Her research focuses on relations between culture, design and technology, specifically on social change and transformation through emerging digital technologies. Exploring and developing theoretical and methodological approaches of research in and through design, between social anthropology, participatory design and interaction design, her work has contributed to the development of design anthropology as a transdisciplinary field of academic research. She is co-founder of the International Research Network for Design Anthropology and co-editor of two books at the forefront of design anthropology: *Design Anthropology: Theory and Practice* (2013) and *Design Anthropological Futures* (2016).

Yolande Strengers is Associate Professor of Digital Technology and Society in the Emerging Technologies Research Lab at Monash University. She is also Associate Dean, Equity Diversity and Inclusion at the Faculty of Information Technology, an Associate Director of the Monash Energy Institute and Lead for the Energy Futures research program in the Emerging Technologies Research Lab. She is a digital sociologist and human-computer interaction design scholar who specialises in understanding the gender and sustainability effects of emerging technologies in the home. She is co-author of *The Smart Wife* (MIT Press, 2020), with Jenny Kennedy, and numerous publications exploring the role of automated devices in everyday life.

Jakob Svensson is Professor of Media and Communication Studies at Malmö University, Sweden. He is currently finishing a research project funded by the Swedish Research Council on people and cultures behind algorithms and automated systems. Other research interests revolve around mobile phones and empowerment in the global south, as well as digital media and political communication.

Xue Ying (Jane) Tan is Software Engineer in the Digital Media Research Centre at the Queensland University of Technology. Her research focuses on machine vision to explore everyday promotional cultures in visual social media platforms.

She obtained her Bachelor of Information Technology and Master of Computer Science degrees from the University of Queensland in 2018 and 2019, respectively. Her interests include machine learning, full stack web development and data analytics.

Verity Trott is Lecturer at Monash University and is Member of the Automated Society Working Group in which she investigates the political, cultural and social dimensions of digital technologies. Her research focuses on digital activism, digital culture and data cultures including visualisation practices.

Julia Velkova is Associate Professor of Media and Communication Studies at the Department of Thematic Studies – Technology and Social Change at Linköping University, Sweden. Her research interests include media infrastructures, data materialities and the intersections of digital economies with energy politics, drawing on approaches from STS, cultural studies and media and communication studies. Her work has been published in journals such as *New Media & Society*; *Cultural Studies*; *Media, Culture & Society*; and *Big Data & Society*. She is currently co-editing the book *Media Backends: Critical Studies of the Other Side of the Screen* (2022).

ACKNOWLEDGEMENTS

This volume would not have been possible without the support of various funding bodies. The Swedish foundation Riksbankens Jubileumsfond (grant F19–1053:1) funded the research network Re-humanising Automated Decision-Making that allowed us to come together in Melbourne for a workshop from which this edited volume was developed. We thank the Emerging Technologies Research Lab (ETLab) at Monash University for hosting the workshop, and we are grateful to Bianca Vallentine, ETLab's manager for organising the event. Martin Berg contributed to the editing of this book as part of his research project 'Working with Algorithmic Colleagues: Expectations and Experiences of Automated Decision-Making', funded by the Swedish Research Council (grant number 2020–00977). Deborah Lupton's and Sarah Pink's contributions to editing this book were conducted within their work as members of the ARC Centre of Excellence for Automated Decision-Making and Society (grant ID CE200100005), funded by the Australian Research Council. Minna Ruckenstein's editorial work and the research for the chapter co-authored with Tuukka Lehtiniemi were funded by the Finnish Academy (grant number 332993). Malmö University's strategic research programme Data Society funded open access for this book. We also wish to thank our Routledge editors – Lucy Batrouney and Georgia Oman – for their enthusiasm for this book and their unerring support along the way.

EVERYDAY AUTOMATION

Setting a research agenda

Sarah Pink, Minna Ruckenstein, Martin Berg and Deborah Lupton

Everyday life is increasingly automated with the use of new and emerging digital technologies and systems. Discussion of these automated technologies is often shrouded with narratives which highlight extreme and spectacular examples, rather than the ordinary mundane realities that characterise the overwhelming majority of people's actual encounters with them. When we hear about the practical effects of automation in society, it is usually for one of two corresponding reasons. The first relates to when automated systems go disastrously wrong and receive high levels of public attention. Recent examples include the Australian 'Robodebt' scandal,[1] where an automated system wrongly issued debt notices to vulnerable welfare applicants, and the UK school leavers' exam grading fiasco,[2] where students were sent algorithmically estimated exam grades much lower than those expected. The second reason that automated technologies receive high levels of publicity or promotion is when they have saved, or are predicted to save, lives: for instance, through accident prevention, medical and pharmaceutical interventions or in humanitarian domains.[3]

In contrast, experiences and processes of automation as part of quotidian routines in our everyday lives in our homes, transport, at work and in education have slipped under the radar of much popular and academic attention. *Everyday Automation* brings this domain of our lives into focus through its attention to the mundane. It asks: How, where and when are automated systems and technologies operating and emerging in everyday life across different global regions? What is their likely impact in the present and future? How do engineers, policy makers, industry stakeholders and designers envisage automation as a solution to individual and societal problems? How do these future visions compare with the everyday realities, power relations and social inequalities for which automated systems and technologies are planned? What do people know about automation? And what are their experiences of engaging with 'actually existing' automated technologies?

DOI: 10.4324/9781003170884-1

The contributors to this book discuss these questions in relation to two overlapping manifestations of digitised automation: artificial intelligence (AI) and automated decision-making (ADM). These involve technical systems that are characterised by algorithms. However, all of these terms – AI, ADM and algorithms – escape single definitions and indeed are defined contextually by each contributor, as they discuss their manifestations in particular technologies and fields of practice. An algorithm is 'an abstract, formalised description of a computational procedure' in computer science (Dourish, 2016: 3) or is described more sociologically as instructions for carrying out tasks and solving problems, assembled by professionals and engineering teams (Burrell, 2016). Algorithms are not static; they are always situated, and, as Paul Dourish (2016: 2) puts it, they always 'come to act within broader digital assemblages'.

Moreover, as emerging technologies, definitions of AI and ADM are likely to shift as their capabilities, imagined markets and possible applications change over time. There is no monolithic definition of AI: indeed, there has been a long trajectory of its discussion spanning back to the mid-20th century (Elliott, 2021: 5). The hype, hope and anxiety around the implications of automation in everyday life are commonly centred on the future of AI. Yet, in fact, the subset of automated technologies represented by ADM is already a part of our everyday worlds in both overt and silent ways that receive little attention in public forums. Deborah Lupton's (2021) analysis of references to ADM in the mainstream Australian press from 1997 to 2021 revealed that this term was infrequently used in such outlets. While Lupton found some positive reports on the benefits of ADM technologies, they more often referred to scandals and failures of ADM and portrayed it as 'untrustworthy' or 'inferior to human decision-making'. Indeed, as these concerns grow, researchers and advocacy organisations have become committed to calling out the uses, possibilities and the consequences associated with ADM. This has given rise to a current debate about ADM that covers a variety of technical tools and systems and a plethora of ADM definitions. Designers, legal scholars, policy makers, ethnographers and data scientists tend to rely on incompatible notions of ADM when they discuss the decision-making qualities and possibilities of new and emerging digital technologies.

There are now numerous research institutes, centres, groups and networks globally that focus on the design, development and critique of AI and ADM. We ourselves and many of the authors featured in this edited collection are leaders or members of these research groups.[4] Their very existence and the often considerable levels of funding that have been granted for their establishment demonstrate that there is considerable public, corporate and government interest in and concern about the rise of the latest wave of digital technologies involving AI and ADM, their implications for our futures and how best to regulate them to forestall any harms.

The messiness of the ADM and AI fields might be seen as a problem, and one way forward involves engaging in a cross-disciplinary mapping of ADM and AI definitions to produce taxonomies and classifications for a shared vocabulary. Nevertheless, the cases presented in this book suggest an alternative approach: to depart from the techno-centricity of the debate and define what ADM and AI are contextually after having carefully explored what they do, with whom and to whom. This keeps the focus on both the sorts of social and societal arrangements currently being built with algorithmic systems and technologies, and the kinds of problems they are seen to solve. Above all, in this book, we are interested in how emerging technologies such as AI and ADM participate, or are expected to participate, in cultural and social processes together with humans. *Everyday Automation* argues that to understand the possible present and future of digitised automation, we urgently need a people-focused approach, led by theoretical and methodological approaches of the humanities and social sciences, which acknowledges that people are involved at every stage of the design, delegation and implementation of automated systems and technologies.

Collectively, the contributors to this book demonstrate this by bringing together research developed across anthropology, sociology, media and communication studies and ethnology, which shows how by rehumanising automation – acknowledging the multiple roles that humans play in relation to automated systems and technologies – we can gain deeper understandings of both these technologies' societal impacts and the impacts human have on technologies. The contributors achieve these relational insights through close examinations of ADM and AI in medicine and public health, the smart energy industry, mobilities, marketing and advertising, administration, fashion, smart homes, platform labour, social services and service industries, education and the news media.

Rehumanising automation

AI and ADM systems, technologies and devices do not and cannot exist independently or autonomously from human thought, embodiment and action. They are always inextricable from humans; they are entangled within social relationships, cultural contexts and human-made infrastructures and institutions (Lupton, 2019). As social science and humanities scholars, we need to *rehumanise* automation. The task of rehumanising is not a new research endeavour, but it is an urgent one. With the spread of ADM and AI from media and health to urban planning, homes, work and education, computational procedures shape aspects of the everyday. Credit scoring, hiring practices, allocation of social benefits, social media engagement and healthcare diagnostics now take advantage of ADM and AI. Rehumanising is a starting point for exploring the complexities of AI and ADM systems by establishing the human as a critical and creative agent in sociotechnical transformations and human–machine relationships. A focus on rehumanising allows us to make visible the human discontents, forces and imaginaries in relation to AI and ADM systems as well as surfacing the possibilities generated by these enactments and assemblages.

Existing research reveals two poles that tend to narrow down the current debate. On the one side, industry imaginaries frequently represent automated technologies as complex yet seamless, offering almost magical solutions to problems (Elish and boyd, 2018; Mateescu and Elish, 2019; Dahlgren et al., 2020). On the other side, a counter-imaginary portrays data-driven automated decisions as cruel, inaccurate and reductionist (Lupton, 2021), suggesting technologies will soon make people redundant across a wide array of domains (Brennen et al., 2022; Köstler and Ossewaarde,

2021; Ouchchy et al., 2020). Other critics further position AI, ADM and related technologies as operating to dehumanise people by not properly acknowledging human differences, individual lived experiences, socioeconomic inequalities and cultural contexts in data-driven decision-making. According to this viewpoint, humans are rendered into collections of data points, with their individuality, feelings and embodiment rendered invisible in algorithmic processing that relies on generalisability and simplification of complexity (McQuillan, 2021). As McQuillan (2021: 70) points out: 'There is nothing personal about the predictions of AI – at root, they are always some form of labelling in terms of "people/objects like you"'.

Our emphasis on rehumanising automation is a response to the historically rooted and persistent dichotomised imaginaries of digitised automated devices and systems. We insist on the need to account for how humans are involved in AI and ADM systems and technologies at every stage of their design, development and implementation. These devices and systems involve the engineers and computer scientists who develop and design automated technologies and systems and the organisations that constitute their markets (Seaver, 2019). Supposedly, 'automated' services are regularly propped up by human workers behind the scenes (Mateescu and Elish, 2019). More recently, however, some industry discussion of the merits of the 'AI-Human Hybrid Chatbot'[5] has emerged, and research about older people's experiences of smart home technologies has shown specifically that such devices are usefully combined with human support services (Pink, 2021; Strengers et al., 2021). There is, moreover, a massive academic research enterprise in the creation of AI and ADM, sustained by research funding and involving humans who determine research policy and priorities, frame funding calls, review funding applications and award funds.

All the human decisions that are made in these environments inflect the ways that AI and ADM proceed as part of our lives. Yet the emphasis in existing media coverage, research and industry or advocacy reports often focuses on what automation *does to people* (see Dahlgren et al., 2020; Lupton, 2021), rather than on what *people do with automation.*

Beyond regulatory ethical frameworks

The current problem is not that there is no concern about people in existing research and policy regarding AI and ADM but that this work usually fails to engage with people in their everyday worlds. In fact, considerable debate and critique – some of which we and the contributors to this edited volume have participated in – exist about questions of 'Automating Society'[6] and 'The Algorithmic Society'.[7] Yet much research and strategy directed towards bringing these technologies into society is still seen as being dependent on their ethics, governance and regulation, as if once these issues were sorted out, then they would be able to effectively function in society as semi-autonomous agents. Ongoing developments have created a situation where societally shared values, ranging from trust and solidarity to autonomy and equality, can become compromised with the implementation of new digital infrastructures (Sharon, 2018; Prainsack and Van Hoyweghen,

2020). However, the questions related to values are typically treated narrowly or at a high-theory level, without closely paying attention to what goes on in society: what people actually do and think.

These developments have energised legal, regulatory and ethical approaches, promoting new governance frameworks and debates around privacy, fairness and ethics (Marelli et al., 2020). The recent proposal for harmonised rules on AI in the EU (European Commission, 2021) is an example of a regulatory attempt to navigate the negative societal consequences and potential socioeconomic benefits of ADM systems. In all these well-meaning initiatives, everyday lives remain at best a curious sidenote or at worst ignored altogether.

The ethics associated with automated systems and technologies foregrounded in much of the existing academic literature also frequently diverts attention away from the ethics implied by the situatedness of AI and ADM in the everyday. Instead, dominant approaches have tended to treat the everyday as a landing site upon which these emerging technologies will make an impact. These approaches have consequently argued that AI and ADM technologies must be designed and regulated to make them ethical (on terms defined by experts) prior to being allowed to make an impact on people. For example, prominent legal and science and technology studies (STS) scholars (e.g. Jasanoff, 2016) and ethicists (e.g. Floridi, 2019) have tended to assert that the problem is that AI and other emerging technologies such as ADM and machine learning are being developed first, with regulation, governance and ethics applied as an afterthought, whereas ethics really needs to be considered at the outset. They are right that ethics must be prioritised. But such arguments still do not fully account for people, in that they put regulatory and ethical frameworks before considering the actual practical experiential and social and political encounters, meanings and implications that emerging technologies might have in diverse everyday worlds. They take ethics out of the everyday rather than engaging with it in the everyday.

New governance initiatives, organised around fairness, accountability and transparency are of course important, but they can leave a lot to be desired from the perspectives of the social sciences and humanities if they ignore decades of social scientific research, employing off the shelf, normative or philosophical understandings of values and ethics. Much of the current literature on fairness and trust in fields like human–computer interaction, for instance, locates values within algorithmic operations (fairness as a statistical property of models), ignoring the differing ways that values might be understood in the larger contexts in which algorithmic systems are embedded (Lanzeni and Pink, 2021). To truly promote ethically sustainable automation, approaches are needed which account for values and ethics as emergent and relational, responding to various circumstances of life.

We argue that, instead, for ethics to come first, people need to come first, since ethics and values cannot and should not be separated from people and their everyday lives. In her chapter, Sarah Pink discusses how ethics and trust in AI and ADM have become bound up in industry and government frameworks which treat them as commodities which can be extracted from faceless publics and invested in machines. It is assumed that these machines will subsequently be considered ethical and that people will then invest their trust in them. Pink proposes that a recourse to anthropologies of ethics and trust, which locate both as continually changing everyday feelings, is needed to reorient work on AI and ADM ethics as an interdisciplinary field which accounts for the social sciences.

In business circles, a quest to anticipate and guide industry and policy makers through automated futures has become a key theme in the work of the consultancies and technology companies. A review of technology and energy industry reports on automated home technologies revealed that these reports rarely attend to the complexities of everyday life, relationships or experiences (Dahlgren et al., 2020; Strengers et al., 2022), but the question of how to better design for ethical, fair, accountable, transparent or unbiased automation is frequently raised. Moreover, frameworks for responsible and ethical AI abound. A 2019 review study 'identified 84 documents containing ethical principles or guidelines for AI' (Jobin et al., 2019) from a mix of industry, government and consultancy sources. Yet these frameworks are rarely underpinned by deep understandings of the diverse people in whose everyday lives, relationships and experiences automation is having varied uses and meanings. In fact, such people are not usually considered as active participants in the ethical AI futures that ethics frameworks prescribe.

Ethical considerations of the design or deployment of emerging technologies also often fail to recognise the effects of major social transformations or challenges and how ethical evaluations may change in response. The COVID-19 pandemic is one such global transformation, throwing societies around the world into disarray as they faced not only a health crisis but also serious socioeconomic upheavals. In her chapter, Deborah Lupton shows how during the pandemic, in the name of crisis management, the promissory narratives and practices around the rollout of automated technologies for monitoring and control of the novel coronavirus precluded acknowledgement of the diverse everyday circumstances and inequalities that they sometimes exacerbated. In this case, discussions of ethics were located elsewhere, outside the realm of the everyday. The demands of the crisis trumped considerations of the ethics of managing people's movements and limiting their freedoms with the use of digital devices and software. Together with other restrictions and surveillance imposed by governments and health authorities, people were expected to accept greater personal surveillance and limits on their movements enforced by novel technologies as the trade-off for protecting their own health and that of the body politic. The ethical implications of these restrictions and monitoring systems were rarely openly discussed or debated, with the social licence for their imposition assumed to be upheld by the state of crisis. Lupton's analysis, therefore, highlights the relative, situated and arbitrary nature of ethical considerations of emerging technologies: a perspective that contrasts with the normative, fixed and generalised approach that is often articulated in the AI ethics literature.

One of the key limitations of the contemporary avalanche of technologically and governance-driven assumptions and arguments about the benefits and risks of automation to people, and how to achieve or mitigate them, is not simply that they are deceptive because they appear as if they were solutions to a problem. Rather, it is that they represent singular interpretations and agendas. In fact, it is the inevitable incompleteness of these existing critiques and calls for ethics and regulation from the social sciences and humanities which illuminates the need for an interdisciplinary approach. Approaches that account for the inseparability of people, ethics and technology are increasingly being advanced, and this trend indicates that we need to account for ethics differently. This involves attending to a reality where humans as well as animals and other non-human species are inevitably co-implicated with AI and ADM systems and technologies. In this experienced reality, people and other species are situated beings, who inhabit continually changing environments. It is of no surprise that this complexity is usually not accounted for by the scholars, public bodies and industry stakeholders who are most concerned with regulation, policy and governance because understanding people in their complex habitats requires a different kind of expertise and sensibility.

Approaches to automation and ethics as expressed in more-than-human theory, decolonial theory and Indigenous and First Nations philosophies highlight that humans and objects are never separate from each other. Humans make digital technologies and digital data; digital technologies and data make humans in a continually co-evolving set of relationships (Lupton, 2019). As outlined in the 'AI Decolonial Manyfesto' (2021) (so-named because of the plurality of viewpoints expressed therein), the language used to talk about AI and ethics is typically grounded in the perspectives and assumptions of men, whiteness and wealth. The 'manyfesto' goes on to assert that 'We reject the Western-normative language of "ethical" AI and suggestions of "inclusivity" that do not destabilise current patterns of dominant and address power asymmetries'. The authors argue that most current attempts to consider AI ethics are merely 'tweaks' that do little to properly address the fundamental power asymmetries that are currently structuring of and inherent in AI and related emerging technologies – and, indeed, often serve to 'whitewash' these inequities. Separating feeling and being from knowing, materialities from immaterialities, the social from the technical, is a common approach in Westernised, colonial perspectives on the ethics of AI and ADM (AI Decolonial Manyfesto, 2021).

In this context, moves towards better regulation or eliminating bias from ADM technologies without recognising these broader dimensions and historically grounded contexts of human-technical relations and alternative forms of knowledge are merely papering over the cracks. A turn to the everyday forms part of a response to the relational and situating call of such decolonising narratives. It also in turn calls for closer engagement with the arguments of decolonising scholarship in the design for future studies of everyday automation. In our brief to rehumanise automation, we are interested in ADM that takes place in situated *practices* rather than in the abstract. We offer a grounded perspective on current algorithmic developments, staying close to actual empirical cases and people behind algorithms, what they do when they build, engage with, promote and evaluate algorithmic systems (Ruckenstein and Turunen, 2020). Such an intervention, from the everyday, is moreover needed to balance the way that ethics are conceived in approaches that seek to foreground ethics through regulation.

Situating the power of automation

When studying everyday automation, we need to take a stance in relation to the automation logic that relies on computational functions that standardise life processes to facilitate the appropriation of data, preferably for profit (Andrejevic, 2020). This logic works not only on a global scale but also internally on local populations in different parts of the world. The current global internet empires – the American companies such as Google, Facebook and Amazon along with their Chinese counterparts such as Baidu, Alibaba and Tencent – aim to capture everyday practices and translate them into quantifiable data, to be analysed and used for the generation of profit. Other actors, who control computational functions, include developers of digital platforms, data analytics companies and digital marketers, suggesting that an expanding range of professionals are taking advantage of automation and exploring its potentials. Given the informational asymmetries and economic forces, it is not surprising that ADM technologies are associated with grim and dystopian future predictions.

Critiques that are more specific point out how ADM favours some groups of people at the expense of others, or it is not accurate enough in its predictions. We are not arguing against the ideas that these powers are at play, but suggest that we need to push back on universalising tendencies and not treat power as if it were inseparable from the people who design, use and promote technologies. Critical data and algorithm studies of the global data extracting machinery and its effects become complicit in making and sustaining the very paradigms and logics that they critique, if they do not acknowledge the situatedness of processes of power. The critique reifies the data extracting capabilities of technologies, rather than querying how they operate or paying attention to ethnographic realities that question its argument. As such, it follows its own internal logic, which denies that there is any power in human creativity, or the everyday expertise of people across diverse situations. In these universalising and techno-determinist approaches, power is portrayed as operating from above, with linear effects leading to dystopian conclusions. For instance, Couldry and Meijas (2019: 5) portray the power of data colonialism as 'the capitalization of human life without limit'.

The ordinary citizen is represented as passively in thrall to manipulation and exploitation of the proponents of the digital data economy. Yet, the automation logic is not the same everywhere – nor does it operate with the same kind of intensity on every occasion of use or every geographical location. People can and do resist – and, indeed, they may call for more customisation and personalisation (Lupton, 2019; Ruckenstein and Granroth, 2020) or even an expansion of datafication of their lives so that their needs are better met (Milan and Treré, 2020). If we believe that human life can be limitlessly captured with datafying technologies, we are giving far too much credit to technologies and far too little to the human agencies involved. In separating digital technologies from humans in a combative and oppositional relationship, this approach fails to recognise the idea that humans are always part of technologies, and vice versa.

In order to see what specifically is harmful and problematic in automation, we need to recognise which problematic practices are already in place, which are in the realm of possibility and which are merely techno-determinist responses. The cases that are discussed in this book feature local specificities, underlining that societies have their own power dynamics that shape processes of automation. For instance, with the social benefit systems the country has in place, the harm that was caused to vulnerable people by the Australian Robodebt disaster could not take place in Finland, at least for now. Thus, whereas the automation logic seeks universal effects, local developments suggest that they materialise and are responded to in remarkably different ways. When people and organisations work with concepts such as AI and ADM, they affirm them locally and pave the way for technologised futures. Yet, in these processes, these concepts also develop and transform and become sites of negotiation and tension. We need to understand such negotiations and account for the resilience and creativity of people in the everyday life circumstances where automation is encountered, to ensure that automation works for them and to account for the situatedness of the ethics and priorities through which this occurs.

It is precisely on the everyday always emergent ways of knowing and understanding that come about as people encounter technologies in everyday worlds that this book centres. If we ask what happens when we encounter and imagine automation *in* lived everyday environments, with living people and as part of actual lives, then new stories emerge. The contributors to this book take us into professional and everyday worlds to highlight the possibilities that actually emerge as AI and ADM become everyday realities.

Where is everyday automation?

Everyday automation is not necessarily always visible, noticeable or memorable. While industry reports may make regular reference to ADM (AlgorithmWatch, 2020), people's everyday experiences and discussions do not often bring ADM to the fore. A digital home assistant, for example, may not be recognised as an ADM technology by that name. Even the news media tend not to use the term 'automated decision-making' very often, while terms such as 'robotics' and 'AI' are very commonly employed (Lupton, 2021). And this is even more the case when conventional research about ADM technologies fails to investigate exactly how they do become visible, sensorially or affectively experienced in everyday life situations, relationships, places and processes. Industry perspectives often promote the convenience, ease and comfort that automation should bring to people's lives, as it is left to manage the everyday. Automation of mundane tasks, in ways which are promised to free up the time of the individual while benefiting institutions, organisations or society as a whole, has been found or has been imagined across the various sites of everyday life discussed by the contributors to this book.

In the home, dominant discourses see the backgrounding of automation as an advantage, whereby invisible automation can help people run busy lives in energy- and time-efficient ways. For example, in a 'set and forget' scenario, based on industry assumptions about the future, where smart home technology manages otherwise boring everyday tasks and decisions, Strengers et al. (2022) portray visions of smart hot water and laundry systems which optimise energy use. In their contribution to this volume, Julia Velkova and colleagues discuss the example of smart thermostats designed to track and learn people's preferences and automatically regulate the temperatures of their homes while saving energy. Their case demonstrates how the ADM experiment is built to mediate the interests of residents, energy infrastructure providers and data-driven companies. Here, the ADM system consists of a plethora of relations that require careful balancing. Velkova and her colleagues describe how the ADM system creates new kinds of ties between users, data handling properties and company interests, but at the same time these relations remain hidden. This raises questions about the nature of the experiment at hand, and what all is being tested with it. New relations involved in the ADM system suggest material and societal reconfigurations that call for further engagement.

In their chapter, Tuukka Lehtiniemi and Minna Ruckenstein discuss another case of experimentation: an unusual data labour arrangement in which prisoners label Finnish language data for a local AI firm. In current research, data labour is often seen to accelerate precarity and inequality, but Lehtiniemi and Ruckenstein demonstrate that the Finnish prison data labour case is multifaceted in its aims. In demonstrating what is of value to the different parties involved in the organisational arrangements of AI training, they show how the prison data labour both work with and intervene in political–economic incentives and pressures, such as platformisation and automation. By doing so, the case calls for critical inquiry that is able to hold seemingly contradictory aspects of ADM together without resolving them into a totalising perspective that loses important differences and alternative paths and ends up seeing only techno-deterministic futures.

Domestic life is also increasingly a site of contradictory values that have to do with technologies. The chapters in this volume reveal persistent incompatibilities between humans and their technological companions, including digital assistants that are supposed to help people with everyday tasks. In Horst and Mohammid's contribution, we are presented with a study of how the Amazon Echo Look – a device that uses machine learning and AI to support people in deciding what to wear – invites a certain kind of human–machine interaction which helps people to make everyday choices. Drawing on an interview study with women in both the USA and Trinidad, these authors show how the device's built-in learning model did not allow for the kind of nuanced personalisation that fashion would require. In the chapter authored by Strengers and Kennedy, a similar, albeit more ubiquitous and well-known, domestic technology is discussed: the Amazon Alexa, a device that increasingly uses emotions as the basis of decision-making. Drawing on a variety of sources, the authors go beyond the emotional surface of Alexa to show how this technology is underpinned by a series of human decisions that affect how emotions are defined and categorised, how data are collected and what caring forms of interactions between people and machines should look like.

Automation in the workplace takes many forms in which humans are implicated in different ways. Collectively, the chapters in this book suggest that the more humans are acknowledged and involved in the processes and practices through which ADM and AI at work are acquired, applied and used, the more likely it is that they will become productive coworking technologies. Contributors reveal cases where the rationalising and personalising discourse that frame industry narratives about why and how to automate everyday workplace practices miss the point due to their failure to engage with either how work really fits into the everyday life of the very workers on whose lives they are seeking to make an impact.

In his chapter, Martin Berg discusses how industry ideas underpin the automation of the workplace by exploring two world-leading platforms for process automation. He shows that platforms of this kind require that work and work tasks are imagined and categorised as either creative, and thus meaningful, or repetitive, and thus meaningless and – according to companies in this sector, at least – borderline unworthy of human life. Since this way of framing work builds on a very narrow understanding of the realities of professional life, Berg shows how work automation companies market their products and services through stories that create an imaginary universe in which they make perfect sense.

Stine Lomborg's chapter also discusses how automated workplace monitoring software systems, designed externally to supposedly help workers and organisations, do not necessarily account for the realities of workers' lives, in this case drawing on the empirical example of how workers engaged with a technology for self-tracking at work. Lomborg reminds us how the role that self-tracking plays at the workplace depends on the employee's job description and the organisation in question. While technological solutions of everyday AI at work push for standardisation and optimisation aims, people also continue to shape and appropriate digital systems in the contexts where they operate.

The contribution by Magnus Bergquist and Bertil Rolandsson shows how in contrast to situations where automation has been applied in workplaces where it is intended to increase efficiency and output, amongst the healthcare practitioners who participated in their research, ADM was integrated by the healthcare professionals themselves. The result was a commitment to working with ADM in explorative ways that could support their work, amongst a group of professionals who saw themselves as experts who were involved in the design of automated technologies for the purposes they believed they were relevant.

In professional contexts, the experimentation and testing appear to be a key to whether people see ADM as coming from 'outside' of their workplaces, or whether it is something that they domesticate to solve problems that they are facing. Jakob Svensson's chapter examines algorithmic work in the Swedish 'Daily News' room to show how different professionals, including journalists, editors, marketing people and algorithm developers, negotiated their professional priorities through the algorithm: often at weekly in-person coffee meetings. Here, Svensson's ethnography leads to a critical engagement which overturns some of the techno-determinism of critical data studies. He shows in fact how the algorithm developers moderated the journalists' search for solutions via the algorithm and effectively demonstrates the importance of situating any discussions of algorithmic power within the specificity of the relations and nature of power found in any given everyday context.

In contrast to the active professional involvement with ADM found in the studies discussed by Bergquist and Rolandsson and by Svensson, Neil Selwyn sets the scene for an educational context, where Australian teachers are bypassed in the implementation of technologies. Teachers become observers of how automation, in this case in the form of fairly useless facial recognition technology, is pushed to the classroom with erroneous and outdated assumptions about education and what schools might need. Here, technologies become harmful not because they automate but because they are coupled with a mindset that trivialises and instrumentalises everyday encounters at schools. Selwyn outlines the instrumentalisation as inevitable in the context of current educational policies. Yet his case also demonstrates that teachers are well aware of the ill-fitting nature of technology designs, opening possibilities for a careful analysis of which technologies teachers think might actually support educational aims. This would most likely be something much more sophisticated than the stand-alone technology developed for very narrow tasks presented in Selwyn's chapter, requiring the collaboration and co-evolving of teachers, students and technological possibilities.

Everyday mobilities are similarly framed by dominant narratives about how 'we' will travel in the future, which again neglect the realities of the everyday and do not attend to diversity or inclusivity. This includes, for instance, the proposed future scenarios of commuting, which Sarah Pink writes of in her chapter, where your electric car can be automatically wirelessly charged at opportune moments for the energy grid and seamlessly paid through trusted blockchain transactions. In their contribution, Vaike Fors and colleagues discuss the future of autonomous vehicles, envisaged as a solution to the (sometimes imaginary) 'first and last mile' problem of getting people between their homes, transport hubs and places of work. Yet again, these visions of futures have little to do with the ways in which people actually anticipate charging their electric vehicles in the future, or with how they prefer to experience the first and last miles of their commutes.

Knowing where everyday automation is and what it's doing is important for reasons of ethics and responsibility. This is urgent in the private sector and has also been demonstrated in the two public sector examples of the Australian Robodebt and UK GCSE exam grading scandals mentioned at the beginning of this introduction, where automation remains invisible until things break down. Here, uses of everyday automation apparently brought ease to the work-administering systems that are responsible for life-changing information for people living in poverty or the careers of young people. Would these systems have been thrown into the limelight had they brought outcomes that centred the positive wellbeing, nurturing and sustenance of the people whose lives they intervened? Or put differently, had these systems already been visible, transparent and easily accessible and responsible to the people whose lives in which they were implicated, would they have been likely to go so badly wrong? This is a topic that Mark Andrejevic and colleagues discuss in their contribution to this volume. Drawing on an analysis of advertisements on Facebook, they explore methods to produce knowledge about what is going on beyond the promises of platform customisation and individualisation. Methods of the kind they develop can be used to gain a better understanding of this realm and to support discussions about responsibility and accountability. We cannot afford to simply wait for cases of automation going wrong in order to critically deconstruct them. Rather, as the contributors to this book collectively demonstrate, we need to seek out how automation actually plays out across multiple and diverse everyday circumstances and to understand the complexities and contingencies of the dynamics through which it becomes part of life.

The anticipatory modes of everyday automation

There is much to learn from the recent examples of how AI and ADM technologies and systems are entering everyday life discussed by the contributors to this book and highlighted in the previous section. The debates surrounding automation not only are concerned with what is currently underway but also involve anticipatory modes and imaginaries, through which ADM and AI technologies are portrayed as being part of possible futures. There is a rich literature detailing what Sheila Jasanoff (2015) has called 'sociotechnical imaginaries', which in the case of emerging technologies invites us to deconstruct the narratives through which the promise of these technologies and their perceived implications for and impacts with and through the people who design, build and use them are constituted. Identifying and deconstructing dominant promissory narratives about automation invites us to respond through our ethnographic accounts of how automation is playing out in the present in 'actually existing' contexts of use. As mentioned earlier, such analyses make visible, contest and complicate the technologically determinist and solutionist visions of industry, policy and other institutions, while sometimes sustaining their myths in everyday discourses. The constitution of sociotechnical imaginaries is also by definition an anticipatory practice: it is always concerned with predicting or postulating possible futures.

Many of the empirical cases with which the contributors to this book have engaged are supported by the anticipatory stance. People buy into and anticipate and boost the promise of AI and ADM technologies, particularly when they can be seen to be fulfilling some of that promise: such as in the cases of diagnosing rare diseases, creating thermally soothing living spaces or freeing people of the tedious tasks at the office. Optimistic predictions or claims about how emerging technologies will improve 'our' futures are reinforced by speculations of how lives become more fulfilling as these technologies support an unprecedented convenience. A key approach here is to closely engage with cutting-edge technology developments to better understand how they position people in relation to technologies. For instance, the Finnish data activism initiative MyData, a technologically driven effort to rehumanise the digital environment by means of new data arrangements, seeks to promote 'human-centric' data arrangements (Lehtiniemi and Ruckenstein, 2019). Yet human centricity tends to translate into development aims by which humans are efficiently tied to human–technology loops. While aiming to rehumanise, technology development has the tendency to reduce the human, as it expects humans to fit into certain prescribed machinic loops and standardised categories.

The pessimistic scenarios identified in some of this book's chapters are fed by another kind of anticipatory stance. They treat algorithmic systems as external forces that threaten humans and even humanity itself. Here, as Lina Rahm and Anne Kaun's chapter shows, it is also relevant to attend to history. Historically rooted future scenarios that materialise as enthusiasm and anxiety for technologies fail to account for the complex and contingent ways that humans experience, promote and practice future anticipations. Everyday modes of anticipation are characterised by much more nuanced anticipatory sentiments and visions characterised by hope, trust and ambivalence (Pink et al., 2018). These feelings may align with both optimistic and pessimistic narratives, but they also contaminate and contradict the much too clean and purified future visions that dominate industry and policy narratives (Dahlgren et al., 2020; Strengers et al., 2021).

What becomes clear when futures are considered from the social sciences is that imaginaries of sociotechnical futures are likely to be just as messy (Dourish and Bell, 2011), contingent (Bessire and Bond, 2014; Pink and Salazar, 2017) and uncertain (Akama et al., 2018) as those portrayed in the past (as Lina Rahm and Anne Kaun show in their chapter) or in the present. We cannot predetermine the future of automated technologies. However, by examining how they are bound up with the ways futures are imagined, predicted and planned for, as social researchers we can begin to consider how to respond to, critique and intervene in the narratives of future that are predominant. As this book reveals, AI and ADM are not static or simple concepts. Earlier in this introduction, we have pointed out that they are likely to shift and change over time. As such, while AI and ADM themselves are technical research fields, and practical technologies, they come into being in the world in the very technologies that are discussed in the different chapters of this book: in digital voice assistants, work planning and monitoring systems, workplace algorithms, self-driving cars, health and medical technologies, electric vehicle charging, temperature control systems and many more.

Moreover, emerging technologies such as AI and ADM serve as what Sarah Pink has elsewhere called 'anticipatory infrastructures' (Pink et al., 2022) through which to contemplate the future everyday technologies and applications that they make possible and, indeed, imaginable. Here, following the anthropological notion of infrastructures as themselves fluid and relational, while also enabling other things (Larkin, 2013), as infrastructures AI and ADM can be seen as anticipatory devices:

> [I]n being the conveyers of possibility, infrastructures are not only about what might happen in the present, but because they have an inevitable association with the realm of possibility, they are also anticipatory structures, that is they are associated with what might happen next.
>
> *(Pink et al., 2022)*

If we could not conceive of AI or ADM, then we would not be able to imagine the capabilities of the technologies that depend on them, such as those discussed by the contributors to this book: self-driving cars, digital home or workplace assistants, COVID surveillance devices or automated energy systems. We would also not be able to think of more distant mundane technologies, such as blockchain-enabled wireless electric car charging. Subsequently, as future AI and ADM capabilities are developed (and indeed newly define AI and ADM), new sociotechnical imaginaries will be built on them. Given the current state of play in scholarship about automated technologies, this will lead on to a new spate of critical responses to the futures they imply.

As the chapters of this book show, responding to existing sociotechnical imaginaries, as well as actual applications of ADM and AI from the everyday sites in which they are used or implicated, complicates their agendas. Such an approach can also be applied by casting everyday future imaginaries in relief with the sociotechnical visions of dominant narratives. However, the chapters also suggest another starting point, where the everyday is more than a site of response to the colonising tendencies of engineering and science, but a leading collaborator in shaping our present and futures. We return to this point as we define a strategy for future research to conclude this introduction.

A research agenda for everyday automation

The contributions to this book collectively show that everyday automation is a research field in its own right. Yet it is more than just that. It offers a key lesson in understanding automation from the ground up, from the sites where it really plays a role in people's lives and emerging futures. Viewing emerging technologies from any other perspective only touches the surface of their real harms and possibilities because it attends only to the kinds of sociological structures that obscure people and how they feel, imagine, hope and fear. As the examples discussed by the contributors to this book make clear, automated systems and technologies, informed by different intentionalities, manifest and are experienced and engaged with by people very differently in diverse everyday sites, which can be configured in relation to varying relations of power and societal structures.

A new research agenda that foregrounds everyday automation calls for innovative research methods and strategies. The examples discussed by the contributors in the following chapters unfold how in different ways, we might gain entry into the everyday sites where AI and ADM technologies are experienced and imagined. This involves critically immersing ourselves in and following the threads of dominant visions and imaginaries of automated futures; figuring out what ADM and AI are actually *doing* through analyses of, for instance, the ways that they make certain materials and options available to people as they navigate social media or automated energy systems, and how automation platform providers imagine the future of work. Our contributors discuss immersive ethnographic studies in the places where people are already living with AI and ADM; remote or distance digital ethnographies that enable access to the everyday digital routines and places of people which would be complicated to participate in in-person and in-the-moment. Several chapters discuss the value of inhabiting sites of simulation and technology testing where imagined or possible futures are experimented with through speculative and interventional design ethnographies, and where, from an STS perspective, emerging technologies are seen as the experiment. Importantly, a number of the contributors to this volume compare the clean and optimistic future visions of ADM and AI proposed and predicted by industry, government and technology company stakeholders – revealed through analyses of their reports, websites, marketing, news reporting and other materials – with the fine-grained evidence of ethnographies of the appearance of automated technologies in everyday. The results consistently advise us that the situated mundane experience tells us very different stories about the possible futures of AI and ADM, which need to be listened to.

Focusing on what ADM and AI do and how this is made possible by humans, rather than what they are, or are supposed to be, suggests that in order to rehumanise the field, we need longitudinal and situational ADM and AI studies that offer more dynamic and processual views of sociotechnical developments. Historicising ADM and AI, as Lina Rahm and Anne Kaun's chapter shows us, allows us to discern continuity and change in their development over a longer temporal scale. It also enables us to understand the different cultural aims or governance regimes under which automated systems in different sectors and domains have developed and are developing. With a historical sensibility, we can witness the strengthening of existing infrastructures and early efforts to build new ones. This is one way to advance ADM and AI studies: to focus on the different infrastructural arrangements, including the stakeholders involved in the building of ADM and AI systems, and their present or future uses. This means conceptualising and researching in circumstances where ADM and AI are present, rather than analysing their supposed effects from the outside.

The notion of the experiment offers us a way to conceptualise the ways that emerging ADM and AI technologies are becoming part of everyday worlds. There is continuing experimentation with algorithmic systems, in fields from energy to mobility and health to security – both through testing and trialling and through their actual applications in both consumer markets and policy initiatives. This can be conceptualised as a societal experiment, simultaneously taking place in different parts of the world, in very different kinds of societies. We are participants in this global living lab whether we like it or not. Taking an all-knowing position in terms of what the results of this experiment – good or bad – will be has the effect of narrowing the perspective and locking out possibilities to engage with the many alternative futures that can still be crafted. Thus, we need to take a more open-ended stance to the reconfigurations of societies brought about by the testing and experimentation in order to avoid absenting ourselves from the futures being made.

To achieve these purposes, empirical analyses of actual cases are essential, as they can foreground who plans, designs, implements, uses and repairs automated systems and thereby rehumanise the study of automation. Alongside the design intent, attention should be paid to the changing nature of ADM and AI systems over time, as these systems continue to develop with their implementations and uses. Many ADM and AI systems are 'permanently beta', meaning that they are never complete or finished products to be launched into predefined markets, but rather they are constantly in development. This indeed also creates an opportunity to extend the field of everyday automation research into considered interdisciplinary collaborations, which might shift the basis of how AI and ADM are manifested in everyday life technologies and imagined futures towards shared processes and visions that are attentive to the emerging worlds in which they are situated. Empirical everyday life studies enable us to clarify the debates that rage between the techno-optimists and pessimists who quarrel about the effects of AI or ADM. The prism of the everyday highlights how the legal, political and ethical tensions, struggles and consequences of automation actually play a role in people's visions and practices. Immersion in the flow of everyday life brings to the fore the continually changing nature of human-experienced realities, emphasising that relations between people and AI and ADM are similarly always in flux.

To end, we reiterate that in order to renew the conversation, ADM and AI debates need to let go of two key assumptions that have underpinned existing critiques: first, the techno-centricity that treats ADM and AI as stand-alone products, innovations or solutions to existing infrastructural inefficiencies and gaps; second, the critical discourse that treats technologies as a 'general' threat and in doing so makes itself unintentionally complicit in the former narrative by endorsing its techno-determinist underpinnings. Instead, ADM and AI need to be treated as complex sociotechnical systems that develop over time and need ongoing stabilisation, repair and care of human-algorithm relations within the mundane everyday worlds of all the humans who are co-implicated with them.

Notes

1 https://theconversation.com/robodebt-was-a-fiasco-with-a-cost-we-have-yet-to-fully-appreciate-150169
2 https://blogs.lse.ac.uk/impactofsocialsciences/2020/08/26/fk-the-algorithm-what-the-world-can-learn-from-the-uks-a-level-grading-fiasco/
3 www.forbes.com/sites/serenitygibbons/2020/08/25/5-life-saving-applications-of-artificial-intelligence/?sh=59b9ea8b1c58
4 Sarah Pink, Deborah Lupton, Mark Andrejevic and Vaike Fors are all members of the Australian Research Council Centre of Excellence for Automated Decision-Making & Society (grant ID CE2001000005) www.admscentre.org.au/; Martin Berg, Minna Ruckenstein, Deborah Lupton, Sarah Pink, Vaike Fors, Magnus Bergquist, Bertil Rolandsson, Jakob Svensson, Julia Velkova and Rachel Charlotte Smith are all members of the Swedish Foundation Riksbankens Jubileumsfond's (RJ) Re-humanising Automated Decision-Making Network https://mau.se/forskning/projekt/re-humanising-automated-decision-making/. Martin Berg and Jakob Svensson (and Sarah Pink as a member of the advisory board) are members of Malmö University's strategic research program Data Society http://mau.se/datasociety. Minna Ruckenstein collaborates with Algorithm-Watch, and is a director of a Finnish Academy funded project on rehumanising ADM, with Martin Berg, Deborah Lupton and Sarah Pink as members of its advisory board.
5 https://techsee.me/blog/customer-service-chatbot-human-hybrid/
6 https://automatingsociety.algorithmwatch.org/

7   www.routledge.com/The-Algorithmic-Society-Technology-Power-and-Knowledge/
    Schuilenburg-Peeters/p/book/9780367204310

References

AI Decolonial Manyfesto (2021) Available at: https://manyfesto.ai/index.html

Akama Y, Pink S and Sumartojo S (2018) *Uncertainty and Possibility: New Approaches to Future Making in Design Anthropology*. London: Bloomsbury.

AlgorithmWatch (2020) Automated Decision-Making Systems in the COVID-19 Pandemic: A European Perspective. Available at: https://automatingsociety.algorithmwatch.org/

Andrejevic M (2020) *Automated Media*. London and New York: Routledge.

Bessire L and Bond D (2014) Ontological Anthropology and the Deferral of Critique. *American Ethnologist* 41(3): 440–56. https://doi.org/10.1111/amet.12083.

Brennen JS, Howard PN and Nielsen RK (2022) What to Expect When You're Expecting Robots: Futures, Expectations, and Pseudo-artificial General Intelligence in UK News. *Journalism* 23(1): 22–38. https://doi.org/10.1177/1464884920947535.

Burrell J (2016) How the Machine 'Thinks': Understanding Opacity in Machine Learning Algorithms. *Big Data & Society* 3(1). https://doi.org/ 10.1177/2053951715622512.

Couldry N and Meijas A (2019) *The Costs of Connection: How Data Is Colonizing Human Life and Appropriating It for Capitalism*. Stanford, CA: Stanford University Press.

Dahlgren K, Strengers Y, Pink S, Nicholls L and Sadowski J (2020) *Digital Energy Futures: Review of Industry Trends, Visions and Scenarios for the Home*. Emerging Technologies Research Lab. Monash University.

Dourish P (2016) Algorithms and Their Others: Algorithmic Culture in Context. *Big Data & Society*. https://doi.org/10.1177/2053951716665128

Dourish P and Bell G (2011) *Divining a Digital Future: Mess and Mythology in Ubiquitous Computing*. Cambridge, MA: MIT Press.

Elish MC and boyd d (2018) Situating Methods in the Magic of Big Data and AI. *Communication Monographs* 85(1): 57–80.

Elliot A (2021) *Making Sense of AI: Our Algorithmic World*. Cambridge: Polity.

European Commission (2021) Proposal for a Regulation Laying Down Harmonised Rules on Artificial Intelligence (Artificial Intelligence Act). 2021/0106/COD.

Floridi L (2019) Establishing the Rules for Building Trustworthy AI. *Nature Machine Intelligence* 1: 261–2. https://doi.org/10.1038/s42256-019-0055-y

Jasanoff S (2015) Future Imperfect: Science, Technology, and the Imaginations of Modernity. In: Jasanoff S and Kim S-H (eds) *Dreamscapes of Modernity: Sociotechnical Imaginaries and the Fabrication of Power*. Chicago, IL: University of Chicago Press, 1–33.

Jasanoff S (2016) *The Ethics of Invention*. New York: W. W. Norton & Company.

Jobin A, Ienca M and Vayena E (2019) The Global Landscape of AI Ethics Guidelines. *Nature Machine Intelligence* 1: 389–99. https://doi.org/10.1038/s42256-019-0088-2

Köstler L and Ossewaarde R (2021) The Making of AI Society: AI Futures Frames in German Political and Media Discourses. *AI & Society*. https://doi.org/10.1007/s00146-021-01161-9

Lanzeni D and Pink S (2021) Digital Material Value: Designing Emerging Technologies. *New Media & Society* 23(4): 766–79. https://doi.org/10.1177/1461444820954193

Larkin B (2013) The Politics and Poetics of Infrastructure. *Annual Review of Anthropology* 42(1): 327–43.

Lehtiniemi T and Ruckenstein M (2019) The Social Imaginaries of Data Activism. *Big Data & Society* 6(1). https://doi.org/ 10.1177/2053951718821146

Lupton D (2019) *Data Selves: More-Than-Human Perspectives*. Cambridge: Polity.

Lupton D (2021) 'Flawed', 'Cruel' and 'Irresponsible': The Framing of Automated Decision-making Technologies in the Australian Press. *Social Science Research Network*. Available at: https://papers.ssrn.com/sol3/papers.cfm?abstract_id=3828952

Marelli L, Lievevrouw E and Van Hoyweghen I (2020) Fit for Purpose? The GDPR and the Governance of European Digital Health. *Policy Studies*. Available at: https://doi.org/10.1080/01442872.2020.1724929

Mateescu A and Elish MC (2019) *AI in Context: The Labor of Integrating New Technologies*. New York: Data & Society Institute.

McQuillan D (2021) Post-humanism, Mutual Aid. In: Verdegem P (ed) *AI for Everyone? Critical Perspectives*. London: University of Westminster Press, 67–83.

Milan S and Treré E (2020) The Rise of the Data Poor: The COVID-19 Pandemic Seen From the Margins. *Social Media + Society* 6. Available at: https://doi.org/10.1177/2056305120948233

Ouchchy L, Coin A and Dubljević V (2020) AI in the Headlines: The Portrayal of the Ethical Issues of Artificial Intelligence in the Media. *AI & Society* 35(4): 927–36.

Pink S (Director) (2021) *Smart Homes for Seniors*. Documentary film 32.5 mins. Emerging Technologies Lab. Monash University.

Pink S, Dahlgren K, Strengers Y and Nicholls L (2022) Anticipatory Infrastructures, Emerging Technologies and Visions of Energy Futures. In: Valkonen J, Kinnunen V, Huilaja H and Loikkanen T (eds) *Infrastructural Being: A Naturecultural Approach*. Basingstoke, UK: Palgrave Macmillan.

Pink S, Lanzeni D and Horst H (2018) Data Anxieties: Finding Trust and Hope in Digital Mess. *Big Data and Society* 5(1). https://doi.org/10.1177/2053951718756685

Pink S and Salazar JF (2017) Anthropologies and Futures: Setting the Agenda. In: Salazar J, Pink S, Irving A and Sjoberg J (eds) *Anthropologies and Futures*. Oxford: Bloomsbury, 3–22.

Prainsack B and Van Hoyweghen I (2020) Shifting Solidarities: Personalisation in Insurance and Medicine. In: Van Hoyweghen I, Pulignano V and Meyers G (eds) *Shifting Solidarities. Trends and Developments in European Societies*. Palgrave Macmillan, 127–51.

Ruckenstein M and Granroth J (2020) Algorithms, Advertising and the Intimacy of Surveillance. *Journal of Cultural Economy* 13(1): 12–24.

Ruckenstein M and Turunen LLM (2020) Re-humanizing the Platform: Content Moderators and the Logic of Care. *New Media & Society* 22(6): 1026–42.

Seaver N (2019) Captivating Algorithms: Recommender Systems as Traps. *Journal of Material Culture* 24(4): 421–36.

Sharon T (2018) When Digital Health Meets Digital Capitalism, How Many Common Goods Are at Stake? *Big Data & Society* 5(2). https://doi.org/ 10.1177/2053951718819032

Strengers Y, Dahlgren K, Pink S, Sadowski J and Nicholls L (2022) Digital Technology and Energy Imaginaries of Future Home Life: Comic-strip Scenarios as a Method to Disrupt Energy Industry Futures. *Energy Research and Social Science* 84: 102366.

PART I

Challenging dominant narratives of automation

1

IMAGINING MUNDANE AUTOMATION

Historical trajectories of meaning-making around technological change

Lina Rahm and Anne Kaun

Introduction

The current wave of automation, spurred by developments in artificial intelligence (AI), has been described as the second machine age (Brynjolfsson and McAfee, 2014) and the fourth industrial revolution (Schwab, 2017). One important part of this new era of smart machines is large-scale automation, not only of industrial production but also, and more importantly, of our everyday lives. Smart devices and the internet of things are supposed to make our lives, including our homes, smoother and more efficient. The historical descriptions of these tremendous changes often depict a linear development from steam-powered industrialisation and mass production (which also includes the invention of the railway and mass transportation of the first machine age) to large-scale digitalisation with the help of computers that is often depicted as replacing cognitive power. As Brynjolfsson and McAfee argue, what steam power was for the industrial age, the computer is for the second machine age.

This chapter aims to critique histories of automation that draw a picture of technological development as a teleological movement from industrial automation to 'smart' machines, moving from the automation of manual tasks to automating cognitive labour. Instead, we demonstrate that technological innovation is never straightforward but characterised by failures and dead ends as well as specific choices that are anchored in the social and political contexts rather than a natural evolution towards the 'best' technological solutions. Drawing on visualisations of automation in Swedish mainstream press since the 1950s, we focus on critical junctures of automation – such moments where it becomes apparent that automation develops into a different direction than initially imagined. By drawing on these materials, we emphasise the importance of mundane ways of imagining technological change as a way of meaning-making.

DOI: 10.4324/9781003170884-3

Imagining automation

In his seminal book *Forces of Production*, author David Noble (2011) develops a social history of industrial automation and argues that imaginaries are an important part of technological development. Technological imaginaries are, according to Noble, performative, brought to bear as engineers envisage technological development in many ways; for example, in terms of what is possible to automate and what is not. This has consequences for concrete technological development. Noble gives the following example:

> [I]f an engineer was to come up with a design for a new technical system which required for its optimal functioning considerable control over the behavior of his [sic] fellow engineers in the laboratory, the design would perhaps be dismissed as ridiculous, however elegant and up-to-date its components. But if the same engineer created the same system for an industrial manager or the AirForce and required, for its successful functioning, control over the behavior of industrial workers or soldiers, the design might be deemed viable, even downright ingenious.
>
> *(Noble, 2011: 44)*

These possibilities for imagining different future trajectories for technological development are strongly linked to questions of power in society, he argues further. Engineers and designers define to a large extent what is possible at the same time as they

> rely upon their ties to power because it is the access to that power, with its huge resources, that allows them to dream, the assumption of that power that encourages them to dream in an expansive fashion, and the reality of that power that brings their dreams to life.
>
> *(Noble, 2011: 44)*

Technological history often focuses on the paths taken and describes them as naturally evolving, not as situated and partly contingent choices that are based on power relations. Often, mundane technologies appear as the inevitable result of a development path that logically had to be taken. Here, Noble draws a parallel between technological development and natural selection; only the best and most successful technologies will make it, so the dominant narrative goes. In that sense, technological development is imagined as autonomous and neutral, a rational and self-selecting process. This neglects the involvement of people and practices in this process that are far messier and more complicated than often depicted in histories of technological development (Noble, 2011). Noble, therefore, suggests also considering 'paths not taken' as important ways to historicise contemporary forms of automation and technologies on which automation is based.

Material and analytical approach

The material on which the chapter is based is gathered from one of Sweden's largest press clipping archives at the Sigtuna foundation. All newspaper clippings, which have been archived since the beginning of the 1900s until 2000, are sorted chronologically in envelopes by topic. The clippings have been collected from all the Swedish newspapers, as well as the largest newspaper in the neighbouring countries such as Finland, Denmark and Norway, including evening papers such as *Aftonbladet* and *Kvällsposten*, as well as quality national newspapers such as *Dagens nyheter* and *Svenska dagbladet*. As Johan Jarlbrink (2010) has argued, the Sigtuna clipping archive aimed to capture a broad picture of contemporary debates and favoured longer articles and essays over short daily updates on local events. Accordingly, the archival material not only represents a very specific selection of news but also effectively illustrates broad discourses in a specific time period.

We systematically searched the catalogue that is organised around keywords to identify relevant articles and visualisations. The press clippings themselves are not digitised or directly searchable. Hence, we systematically worked through the collections of clippings labelled 'social issues' (including work environment and working time, unemployment issues and leisure), 'technology' (including rationalisation of domestic work, technology from a social point of view, automation, operations analysis, electronics, history of technology, technical research, cybernetics and computing), 'crafts and industry' (including computer industry and graphic industry) and 'economics' (including machine culture, rationalisation, time studies, standardisation and technocracy) to identify relevant images. After having identified a set of relevant images, we returned to the digital archive of Swedish newspapers hosted by the Royal Library to extend the search with keywords based on our first findings in the clipping archive.

The advantage of working with this analogue clipping archive is that the search is broader and less dependent on changing terminologies. As digital development is characterised by its constant emphasis on being new and revolutionary, it follows that a key way to describe technology as new is through the way it is named. Hence, throughout history, digital media technologies have been constantly re-conceptualised. Consequently, as digital media technologies evolve, they often change names to illustrate their 'newness' and to distance themselves from older technologies. Terms such as automation, robotics, electronic brains, cybernetics, computers, mathematical machines, electronic data processing (EDP or ADP), AI, home computer, personal computer (PC), information technology (IT), information and communication technology (ICT) and algorithms partly denote different technologies in different time periods. However, they also overlap both technically and conceptually. By systematically reviewing press clippings, we aim to make emerging ways of thinking about and making sense of the computerisation and hence automation of society visible. We were, hence, exposed to far more articles than we would have been with keyword-based search in a database of newspapers.

The representations that we zoom in on here are emerging around two critical junctures: namely, the first automation backlash in the late 1960s and then the increased critique of data-based automation in the 1970s and 1980s. The two critical junctures in the automation of everyday life since the 1950s in Sweden illustrate how technological development is always also based on 'paths not taken'. Automation in the 1950s was imagined as all-encompassing and concerning not only industrial but also, and in particular, the automation of household chores and everyday life. However, by the beginning of the 1960s, doubts emerged, and articles such as 'Automation – is it really happening?' discussed the extent to which the dream of automation had, in fact, been realised. It also became increasingly clear that decisions to automate certain tasks rather than others were deeply entangled with power relations. Instead of a large-scale automation of the industrial sector, automation happened to a large extent at offices, involving tasks that were conducted by employees with a much weaker union representation than industrial workers at that time. The second critical juncture is an emerging discourse of surveillance and privacy that developed in the late 1970s and early 1980s. Instead of 'dreaming big' in terms of technological change, the popular debate was increasingly dominated by automation anxiety (see also Bassett and Roberts, 2019). Automation was linked not so much to job losses but to questions of integrity and privacy that emerge with information gathering.

Taking these moments and controversies in the automation debate as a starting point, we focus our analysis on mundane ways of imagining technological development: namely, the visual depiction of automation in the popular press. In doing so, we approach technological development by analysing processes of representation as forms of meaning-making. The meaning-making around technology is related not only to how we imagine technology to work with and for us but also to which aspects we ignore in order to develop mundane, workable understandings of complex systems (Star and Bowker, 1999; Gitelman, 2008; Peters, 2015). Robert Willim (2017) has defined this process as mundanisation. With reference to domestication theory (Morley and Silverstone, 1990; Silverstone, 2005) that has conceptualised the integration of new technologies in our everyday lives – the taming of new technology – Willim argues that in order to be able to establish mundane uses of technologies, we have to highlight certain aspects while we forget and ignore others of these complex systems. As such, various lay theories about how technology works become important. We argue that popular media play an important role in conveying such theories: for example, by catering for general audiences rather than experts. We further argue that struggles and conflicting definitions around the meaning of certain technologies render technology and its process of mundanisation visible.

Imaginaries and what they teach us

1950s

In 1955, the Swedish Social Democratic Party and The Swedish Trade Union Confederation organised a now famous conference titled 'Technology and tomorrow's society', where it was argued that, in the future, computers will not only 'relieve the human workforce from heavy and monotonous muscle work, but also from tiring activities in the brain and nervous system' (Velander, 1956: 63). At this conference, the origin of automation is attributed to one of the directors of the Ford Motor Company in America, who during a meeting banged his fists on the table and exclaimed: 'We need more automation!' At the Swedish conference, it was speculated whether 'automation' was a merging of the words automatic and production or perhaps just a mispronunciation of the word automatisation (Velander, 1956). Soon after that, the term 'automation' was further explained as an automatic process controlled by an 'electronic brain' (i.e. computer) and quickly became a viral buzzword of the 1950s (Carlsson, 1999).

The mid-1950s is also the time period where the debate of automation exploded in Swedish popular media (Blomkvist, 1999). Here, it needs to be noted that the Swedish reformist labour movement has had a decisive influence on social development as the Social Democratic Party had uninterrupted government power for 40 years. The conference received a great deal of media attention, and the presentations at the conference were published in book form a year later. This means that early Swedish computer policy is to a large extent also the policy of the reformist labour movement, with a focus on controlling computerisations within the framework of the welfare state (Rahm, 2021). However, the debates in Sweden are not unique; in most countries in the industrial west, the media debate on automation and its potential for societal change began in the mid-1950s.

Automation is, at this point, imagined as revolutionising both the industry and the home, as it was predicted that robots will take over all human work in these domains. As an example, The British Department of Scientific and Industrial Research (1956) proposed that automation would take over virtually all jobs in industry. Despite this, there was also a strong belief that automation would not make workers redundant but rather improve working conditions (Dobinson, 1957). The automated future was expected to increase wealth, reduce the workload, create more leisure for all and thus increase the wellbeing of everyone. As Diebold (1958: 43) stressed in the journal *Computers and Automation*, 'Today we are leaving the pushbutton age and entering an age when the buttons push themselves'.

Figure 1.1 shows an example of how automation was depicted in the 1950s. The caption reads: 'this is what it will look like when the administration is affected by automation'. In a humorous way, automation is presented as a technology for the state apparatus, illustrating how machines were imagined as being in the service of governing institutions, citizens and even humanity at large. Basically, automation is visualised as a rationalisation of existing societal functions, while also remaining quite 'safe' in terms of effects on society. You could argue that such depictions negotiate the function of technology in society by showing its positive and amusing side effects. The underlying plot can be seen as a vision of a 'push-button logic' affecting households, industry and stately governance, where previously complex procedures are now executed via the simple push of a button.

In Figure 1.2, where the text says 'Do not push the button! The machine takes care of itself', we see a further development of the push-button logic. Here, the machine

FIGURE 1.1   The automated state administration, 1956, *TCO-Tidningen/Neue Rhein-Zeitung*.

FIGURE 1.2   Self-regulating automation, 1957, *TCO-Tidningen*.

even pushes the button itself (much as Diebold argued in 1958). The lesson learned from such illustrations is the expectations of a life of leisure from which dirty, dangerous, stressful or threatening aspects are removed. Automated machines will constitute obedient servants to humanity, providing all required services and administration.

Automation is mainly imagined to impact on industry and households, but some effects on offices are also depicted. Illustrations of automation in industry often focus on robots working together with male workers – as a co-worker of sorts. This imagined relationship is effectively shown in Figure 1.3, where an industrial worker is instructing a robot to take it easy when the time efficiency engineer arrives (understood as to not affect piecework ratios). Interestingly, office automation is illustrated differently. In the picture shown in Figure 1.4, women

— Å nu när tidstudiemannen kommer ska du ta det litet lugnt, begrips.

FIGURE 1.3 Automation at work, 1959, *TCO-Tidningen*.

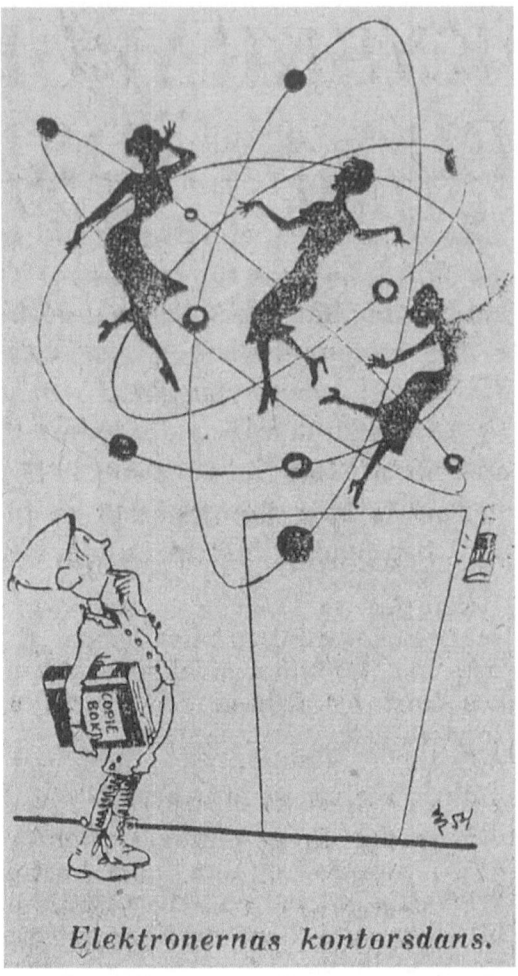

Elektronernas kontorsdans.

FIGURE 1.4 The electrons march into the modern offices, 1954, *Morgon-Tidningen.*

are depicted as forming one unit with the computer – completely immersed in its power – engaged in 'the office dance of the electrons'.

As a finale to the 1950s, Figures 1.5 and 1.6 illustrate how industry and governance are subject not only to automation but also to spare and free time. These pictures are from an article with the headline 'Rapidly emerging technologies create problems with spare time', again illustrating the ideals of the 'push-button logic', in which it is stated: 'Western countries will see increasing problems due to too much spare time, and people will perhaps even start longing for work and job satisfaction' (DN, 25 April 1957). As such, the illustrations of the 1950s seem to prepare citizens for a life where spare time is abundant, where a 'push-button logic of automation' is the norm, and a society where such idleness was imagined as perhaps even becoming a social problem.

*Ring sportstugan, så sätter den själv
på potatisen.*

FIGURE 1.5   Leisure time and automation, 1957, *Dagens Nyheter.*

1960s

The imagined computer revolution of the 1950s did not fulfil its promises – instead of fully automated homes and roboticised industries, it is the state administration that gets computerised at increasing speed, and public authorities, in particular. One indication of this is how the Swedish Trade Union Confederation stated that the computer revolution (and its effects on work) is severely exaggerated and that there is no clear connection between technological change and employment levels. So, instead of the industry being the prime target for automation, it is primarily healthcare, the police, the judiciary, the unemployment offices and educational institutions where computers are seen as providing new efficiency and increased service.

Nevertheless, even though *actual* automation is primarily taking place in state administration, and not within the industry, it is still mainly imagined, and illustrated, as concerning industrial work. The male workers in Figure 1.7 could in an automated future have some alcoholic refreshments during work. We could find no illustrations of female secretaries (or workers) relaxing while a computer is

*Bilarna får autopiloter som kan
kopplas till vid långkörningar.*

FIGURE 1.6 Leisure time and automation, 1957, *Dagens Nyheter.*

doing the main part of the work for them. Instead, as shown in Figure 1.8, female workers in the office setting are shown to a larger extent than the male workers as an integral part of the machine, engaging hands-on in a joint system of human and machine.

In the 1960s, there was also an increase in reports regarding governmental use of computers, mostly focusing on information gathered and stored in various databases. State administration, because of the large quantities of citizen data it deals with, is described as particularly well suited for rationalised computerisation. During this time, computers are often described as neutral systems processing information and depicted accordingly – using rather formal system descriptions. This changed drastically in the 1970s.

1970s

By the end of the 1960s, public discourse was beginning to see the increasing collection of data about citizens by the agencies of the welfare state as a problem.

FIGURE 1.7 'Refreshments may be taken during work' and 'There is no agreement on what automation really is, but still its effect is felt every day'.

Source: Illustration by Dick Stenberg in 1963, *Dagens Nyheter.*

Computers are now increasingly depicted as machines together with humans who are in distress or under control.

In Figure 1.9, it is clear that the tiny human, alone in the shade of the massive computer, is a vulnerable figure. Computing, and in particular its capacity to store and process large amounts of information, is increasingly regarded with scepticism. Popular illustrations at this time often pick up on such themes and promote a view of an exposed individual suffering under an unempathetic computer system of oppression (Figure 1.10).

An increased focus on computerised surveillance means that citizens are described as being undressed at the press of a button. Thus, by the early 1970s, new questions regarding power relations were being highlighted. The article 'Afraid of computers – why is that, then? They will tell everything about everyone' (*Göteborgs Handels- & Sjöfartstidning*, 24 October 1970) is illustrated by two men in suits, each in possession of 'facts' about the other (Figure 1.11).

One of the reasons such issues gained attention was the census and residence registration in 1970 and mainly pertains to the fact that this information is also

FIGURE 1.8   Female automation operators, 1961, *Dagens Nyheter*.

public. As the information was now stored digitally, commercial actors were quick to request data tapes with information about the population. As such, a new market emerges where companies collect and sell the addresses of, for example, all the 'newly divorced 27-year-olds with an income of more than X crowns' or 'all households that purchased a computer last year'. Personalised advertisements emerge as a huge market at the intersection between computerisation and the principle of public access to information. At the same time, computer errors were also gaining attention, and citizens who were subject to glitches suffer unintended major consequences:

> A father of six, who has been hospitalized for pneumonia, received a bill claiming he brought a newborn girl into the world. A woman having an ear operation got a bill for a circumcision. A woman who admits to taking no

FIGURE 1.9 'Computers reflect the society they serve', 1972, Arbetet.

more than five drinks of alcohol per year, is labelled as an alcoholic. Most likely, none of the three who have fallen victim to the feeble mind of the computer, are big fans of electronic brains. But computers are here to stay. As obedient, fast servants, or as reckless masters? Are computers a threat to individual freedom?

(Hufvudstadsbladet, 24/9, 1971)

Computers are described as lightning fast, but once they are fed a false piece of information, the process to correct it is seen as cumbersome. At the same time, government officials are depicted as 'the high priests of computerisation', in exclusive charge over technology, and prone to mistakes that are difficult to remedy – far beyond the control of any layman.

Finally, automation is, at this time, often illustrated using punched cards that were essential for computation at that time – in Figure 1.12, depicted as a pillory keeping a citizen in check.

Demystifying industrial automation in the 1960s

Imagined primarily as an industrial and household revolution, actual automation departs from these visions in the 1960s, and governance and offices are in fact areas

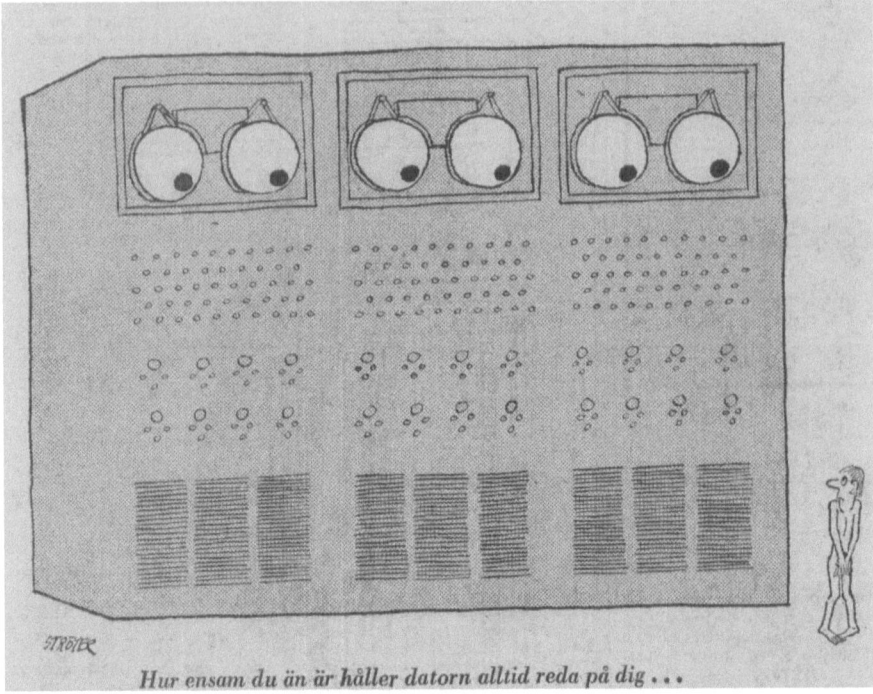

Hur ensam du än är håller datorn alltid reda på dig . . .

FIGURE 1.10 'No matter how isolated you are, the computer keeps track of you'.

Source: From 'Our privacy in databases', 1974, *Dagens Nyheter.*

that are computerised more thoroughly and swiftly. In terms of illustrations and what they teach us, it is clear that the very techno-optimist outlook established in the 1950s, where automation is imagined as a tool in service of society, is beginning to be challenged in the 1960s, where more comprehensive societal systems of control and surveillance, made possible by computerisation and automation, are depicted. In the 1970s, such discourses, imaginaries and illustrations were further reinforced, and automation was seen as controlling and surveilling workers and citizens to a larger extent. The development of the plot (if one was to summarise *one* plot) can be described as a transformation from friendly robot companions towards ubiquitous surveillance systems. The imagined leisure explosion, fuelled by a 'push-button logic', depicted in 1950s illustrations was replaced by the cold gaze of societal control.

The imaginaries of automation from the 1950s to the 1970s included both the robotisation of manual labour (mainly male labour) and 'cognitive automation' (of mainly female labour). By stating this, we want to stress not only that cognitive automation was always part of the overall automation processes but also that much of the labour which underwent early cognitive automation has to a certain extent been overlooked and obscured in public narratives. As such, we can

FIGURE 1.11 'What does he know about me?', 1970, *Göteborgs Handels- & Sjöfartstidning.*

arguably talk of *historic imaginaries of automation* – that is, contemporary visions of history that do not necessarily reflect actual circumstances or processes but rather repeat established narratives and teleologies of particular technological innovations and processes (and thereby obscure certain actors, target groups, technologies or problematisations).

There is also a clear tendency for a form of 'simplified' banal deception when it comes to feminised work. Cognitive automation is not put forward as either advanced work or 'mysterious machinery'. Throughout the studied decades, the image of a woman sitting in front of a computer is constantly repeated. In these pictures, she obviously does more than just press a button, but she is not described as liberated from her work through automation, or as a part of an

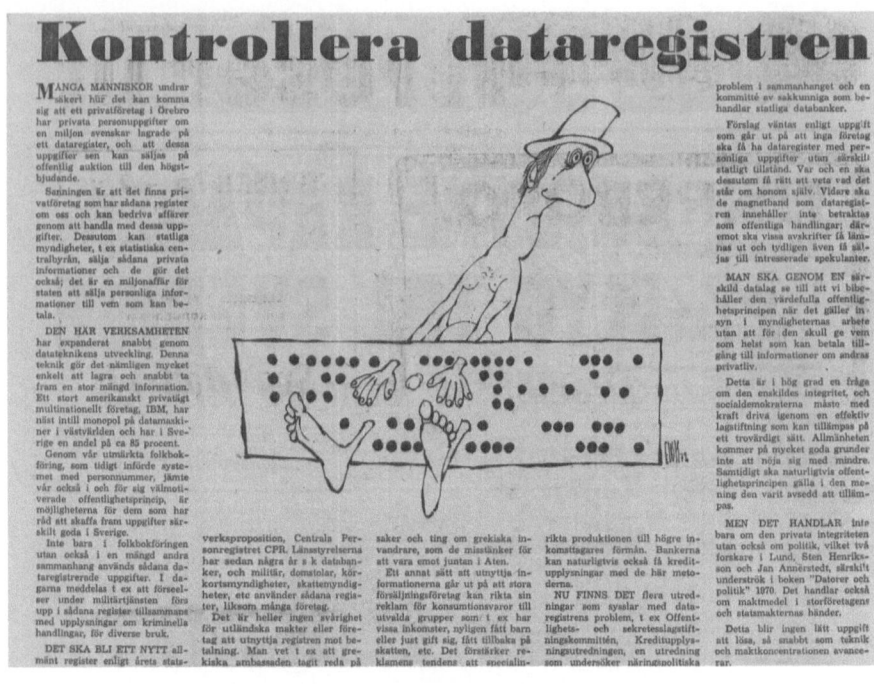

FIGURE 1.12 'Control the databases'.

Source: Illustration by ©Ewert Karlsson (EWK)/Bildupphovsrätt 2021, 1972, in Aftonbladet.

advanced desired technological future (like much of the automation of manual labour is imagined).

Demystifying automation 2.0 in the 1980s

During the second half of the 1980s, Swedish authorities were co-processing data from various databases – something that generated significant debate. The synthesised information is seen as creating severe instruments of control. As an example, at this time the Swedish Agency for Public Management uses data from 300 different databases, mostly maintained not only by public authorities but also by private companies.

Articles and especially their images are an expression of the worries about increased surveillance. For example, an article in the national daily *Dagens Nyheter* illustrates how citizens are turned into data points by the Swedish Social Insurance Agency (which was one of the first public agencies to introduce automated record processing and automated decision-making in the late 1970s). The argument goes that the individual, in all its complexity, is reduced to data and disappears in the computerised system ('We are disappearing in the agency's computer') (Figure 1.13).

Teckning: LARS MELANDER

Vi försvinner i kassans dator

FIGURE 1.13   We are disappearing in the agency's computer by Lars Melander, 1980, in *Dagens Nyheter*.

Similarly, the Swedish Tax Agency is targeted by critique of surveillance. As expressed in this image of an all-watching eye drawn on a new tax declaration sheet (Figure 1.14), this is considered to threaten the integrity of citizens across the country. The all-watching eye of the state as 'Big Brother' is a returning image of the 1980s and gives expression to a general critique of digitised state surveillance.

However, the public discourse also highlighted the benefits of computerisation and automation as dependent on the specific actors implementing these technological changes. The image in Figure 1.15 illustrates this position, arguing that the usefulness and potential of technologies depend on who is trying to tame it. At the same time, computer technology is mystified by the comparison with Aladdin's spirit that serves its master, but which is still a mysterious phenomenon. Computer technology is similarly subservient and mysterious at the same time.

Concluding remarks

Focusing on two critical junctures of digital automation in Sweden, while following the evolving dominant discourse of computerisation and automation

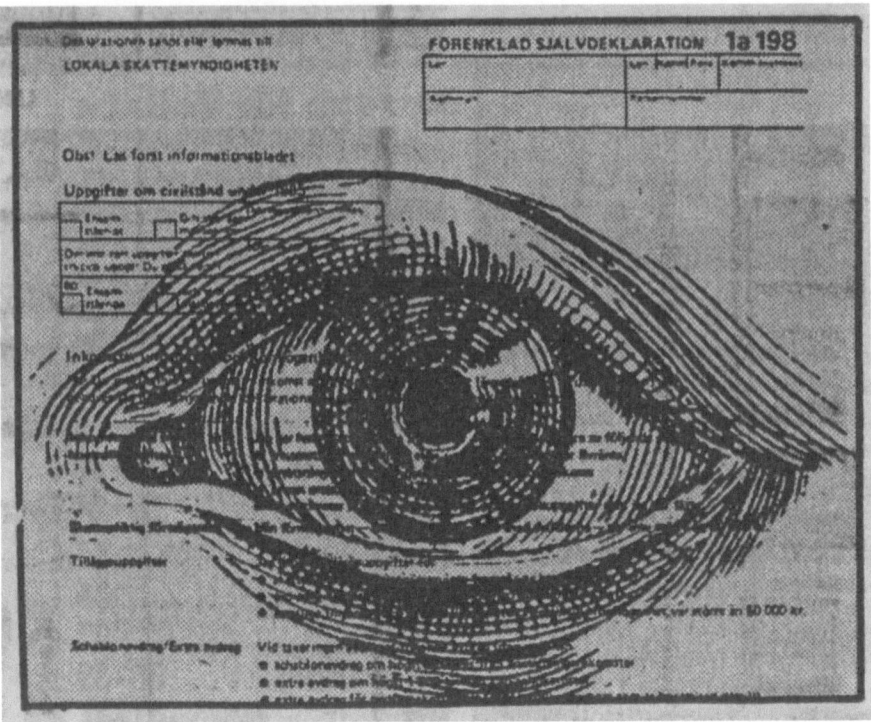

FIGURE 1.14 'Simplified tax declaration threatens our integrity', 1985, *Göteborgs-Posten.*

since the 1950s, we show that an important part of technological change is meaning-making processes that shape and accompany the implementation of new technologies. Sociotechnical imaginaries in the popular press contribute to what Simone Natale (2021) has called make-believe in relation to advanced technologies such as AI. Advanced smart systems only exist to the degree by which we consider them as being smart. However, public discourse and the sociotechnical imaginaries that are negotiated within it are not frictionless. To the contrary, our examples have illustrated that public discourse is moving back and forth between confirming and contributing to the myth production while also critically demystifying technology to a certain degree. The critical demystification of technology remains, however, within a framework of constant technological advancement and automation that is depicted as evolving naturally. The public discourse in the popular press that we have observed is not questioning the premises of automation fundamentally but functions as reconfirmation of the automation imaginary. It is a discourse that develops positions that largely confirm the usefulness and irrevocability of technological change rather than criticising it fundamentally. Hence, it provides the discursive ground for furthering automation in the future.

"*Liksom med anden i Aladdins lampa bestäms datorteknikens karaktär av vem som tämjer den.*"

FIGURE 1.15   'Like the spirit in Aladdin's lamp, the character of computer technology is determined by the one who tames it'.

Source: From 'The Public Bulletin Board', 1988, *Dagens Nyheter.*

Along similar lines, Simone Natale (2021) argues that rather than striving for general intelligence, AI has from the beginning been based on practices of banal deception. AI is a deceitful medium. Only by deceiving us can AI be incorporated into our everyday lives and make its use meaningful (Natale, 2021: 7). Natale argues 'AI only exists to the extent it is perceived as [intelligent] by human users' (2021: 34). In that sense, Natale is interested in the contribution that mundane users make to AI, not in the form of providing data or conducting mundane tasks for further development of specific technologies that others have been engaging with (Irani, 2015; Roberts, 2019; Gray and Suri, 2019) but their contribution in the form of 'make believe' and the acceptance of the deception by AI. These practices of make-believe in the myth, or what Natale calls banal deception, are among a number of preconditions for increased implementation and integration of AI into our everyday lives. This chapter argues that one crucial part of generating and upholding such banal deception can be found in the potentially deceitful narratives and imaginaries of automation in the popular press.

Furthering this argument, Czarniawska and Joerges (2020) claim that there is a dynamic circular relationship between popular discourse and sociotechnical endeavours, where cultural expressions (such as illustrations in newspapers) can influence ideas about technological developments (and vice versa). With reference to Czarniawska and Joerges (2020), our analysis highlights how the functions, plots and time-bound ideas that cultural representations promote are, in fact, *teaching* people about what technology is, can be and should be.

References

Aftonbladet, January 30 (1972).
Bassett C and Roberts B (2019) Automation Now and Then: Automation Fevers, Anxieties and Utopias. *New Formations* 98(98): 9–28.
Blomkvist P (1999) Ny teknik som politisk ideologi. *Arbetarhistoria* 4(92): 13–22.
British Department of Scientific and Industrial Research (1956) *Automation: A Report on the Technical Trends and Their Impact on Management and Labour.* London: Her Majesty's Stationary Office.
Brynjolfsson E and McAfee A (2014) *The Second Machine Age: Work, Progress, and Prosperity in a Time of Brilliant Technologies.* London: W. W. Norton & Company.
Carlsson A (1999) Tekniken – politikens frälsare? *Arbetarhistoria* 4(92): 23–30.
Czarniawska B and Joerges B (2020) *Robotization of Work?: Answers from Popular Culture, Media and Social Sciences.* Northampton, MA and Cheltenham: Edward Elgar Publishing.
Dagens Nyheter, May 13 (1957).
Dagens Nyheter, June 3 (1988).
Dagens Nyheter, August 4 (1974).
Dagens Nyheter, October 1 (1980).
Dagens Nyheter, November 23 (1961).
Dagens Nyheter, December 13 (1963).
Diebold J (1958) Industry and the Automated Future: Problems Along the Way. *Computers and Automation* 6(2): 14–19.

Dobinson CH (1957) The Impact of Automation on Education. *International Review of Education* 3(4): 385–98.

Gitelman L (2008) *Always Already New: Media, History and the Data of Culture.* Boston, MA: MIT Press.

Göteborgs Handels- & Sjöfartstidning, October 24 (1970).

Göteborgs-Posten, March 7 (1985).

Gray ML and Suri S (2019) *Ghost Work: How to Stop Silicon Valley From Building a New Global Underclass.* Boston, MA: Houghton Mifflin Harcourt.

Hufvudstadsbladet, September 9 (1971).

Irani L (2015) Difference and Dependence among Digital Workers: The Case of Amazon Mechanical Turk. *South Atlantic Quarterly* 114(1): 225–34.

Jarlbrink J (2010) The History of the Clippings Archive – The Clippings Archive in History: Clippings in Archive, Diary, and Microfilm [Historiens tidningsklipp – tidningsklipp i historien: Klipp i arkiv, dagbok och bokfilm]. *Historisk Tidskrift* 3: 411–35.

Morgon-tidningen, November 30 (1954).

Morley D and Silverstone R (1990) Domestic Communication – Technologies and Meanings. *Media, Culture & Society* 12(1): 31–55.

Natale S (2021) *Deceitful Media: Artificial Intelligence and Social Life After the Turing Test.* Oxford: Oxford University Press.

Noble D (2011) *Forces of Production: A Social History of Industrial Automation.* New Brunswick, NJ: Transaction Publishers.

Peters JD (2015) *The Marvelous Clouds: Toward a Philosophy of Elemental Media.* Chicago, IL: University of Chicago Press.

Rahm L (2021) Computing the Nordic Way: The Swedish Labor Movement, Computers and Educational Imaginaries From the Post-War Period to the Turn of the Millennium. *Nordic Journal of Educational History* 8(1): 31–58.

Roberts ST (2019) *Behind the Screen: Content Moderation in the Shadows of Social Media.* New Haven, CT: Yale University Press.

Schwab K (2017) *The Fourth Industrial Revolution.* New York: Currency.

Silverstone R (2005) Domesticating Domestication. Reflections on the Life of a Concept. In: Berker T, Hartmann M, Punie T and Ward KJ (eds) *Domestication of Media and Technology.* London: Open University Press, 229–48.

Star SL and Bowker G (1999) *Sorting Things Out: Classification and Its Consequences.* Boston, MA: MIT Press.

TCO-tidningen 10(16) (1956).

TCO-tidningen 11(15) (1957).

TCO-tidningen (8) (1959).

Velander E (1956) Automationen och rationaliseringen. In: Swedish Social Democratic Party & Swedish Trade Union (eds) *Tekniken och morgondagens samhälle.* Stockholm: Tiden.

Willim R (2017) Imperfect Imaginaries: Digitisation, Mundanisation, and the Ungraspable. In: Koch G (ed) *Digitisation: Theories and Concepts for Empirical Cultural Research.* London: Routledge, 53–77.

2

TRUST, ETHICS AND AUTOMATION

Anticipatory imaginaries in everyday life

Sarah Pink

Introduction

The concept of *trust* has come to be associated with an anticipated future in which automated decision-making (ADM) and artificial intelligence (AI) will have been successfully, ethically, inclusively and responsibly implemented in ways that solve societal problems and increase efficiency, safety and quality of life. However, as highlighted by Evgeny Morozov (2013), whose work is well known in this respect, such narratives of technological solutionism which assume that technologically driven societal change will solve social and individual problems are deeply flawed. Critiques of technological solutionist approaches are further advanced by recent work. For instance, Abeda Birhane (2021) brings a relational ethics of Afro-feminism and sub-Saharan African philosophy to bear in a decolonising critique of how Western metaphors and philosophies are mobilised in rationalist problem–solution technology narratives. These works urge us to interrogate the concepts that participate in such narratives and their implications for research and innovation. Moreover, if we are to bring people into the process of ADM technology design, then we need to ensure that they are accounted for in the conceptual categories that frame theory and practice in innovation.

In this chapter, I investigate these questions through the prism of the concept of trust, drawing on the example of transport mobilities, where automation is frequently pitched by industry and policy stakeholders as offering solutions to societal problems. The case of transport mobilities is particularly interesting because it offers an example which has been the subject of enduring and explicit hype and expectation over the last decade. I critically examine the place of trust in the causal logics through which techno-solutionist imaginaries are constituted in this field and the research and innovation paradigms such logics support. I define techno-solutionist approaches to ethics as extractivist, where they seek to identify and

DOI: 10.4324/9781003170884-4

capture human ethics values and invest them in machines with the intention that such ethical machines will engender trust. I argue that, from a design anthropological perspective, a new interdisciplinary approach is needed to achieve ethical automation, which instead accounts for both trust and ethics as contingent, emergent from the everyday worlds and circumstances of life.

The significant step of thinking of automated systems in place of automated technologies is outlined by the non-profit research and advocacy organisation AlgorithmWatch (https://algorithmwatch.org/en/). AlgorithmWatch (which includes AI under the umbrella of ADM) describes ADM systems as 'ways in which a certain technology – which may be far less sophisticated or "intelligent" than deep learning algorithms – is inserted within a decision-making process' (AlgorithmWatch, 2020). ADM systems involve

> a socio-technological framework that encompasses a decision-making model, an algorithm that translates this model into computable code, the data this code uses as an input – either to "learn" from it or to analyse it by applying the model – and the entire political and economic environment surrounding its use.
>
> *(AlgorithmWatch, 2019)*

Visions of ADM and AI, using digital data and algorithms to perform tasks independently of direct human intervention, make them ubiquitous in future transport mobility imaginaries. Here, ADM is sometimes a quiet participant in comparison to its more glamorous counterpart AI. For example, AI has been portrayed as the intelligent technology which will make an ethical decision about who to kill in imaginary autonomous driving (AD) car accident scenarios, where the car is characterised as a 'robot' and subsequently required to fulfil the ethical expectations of robots (Komendantskaya et al., 2021). The techno-solutionist assumption is that if the car was to make the right ethical decisions then publics would trust it. ADM has been discussed as participating in invisible decisions about when to charge an electric car, in the form of an 'automated, machine-learning "decision-assistant"' (Ketter, 2019). In this scenario, as explained later, technology is expected to make the decisions trustworthy. In this chapter, seeing ADM and AI as inseparable from other elements of systems rather than as technological solutions, I explore the implications of these and other examples through attention to the everyday.

Trust has a unique place within techno-solutionist innovation processes. Across numerous publications, podcasts, online news items and statements, engineers, companies and governments claim that when AI and ADM intelligent technologies and systems can be beneficially engaged for 'social good' (e.g. see Tomašev et al., 2020). In these narratives, trust, which is thought to lead to the subsequent acceptance of technology, figures as a requirement for both individuals and society to gain the benefits promised by automation. For instance, these narratives suggest that if people trust AD cars, society will benefit from future efficient, convenient and environmentally sustainable transport technologies and systems. Correspondingly, research into human trust and acceptance of technologies (AD cars are a good example) has been enduringly central to human–computer-interaction (HCI) research (Raats et al., 2020), along with a focus on designing automated technologies that are 'trustworthy' because they display human ethics, fairness and transparency (e.g. Shin, 2021). Getting the ethics right has become a common call out in industry, policy and engineering and computer science disciplines, where it has been suggested that people will trust ADM and AI if human ethics are embedded in their design. However, while these ambitions are well intentioned, they are misaligned with how ethics emerge in the everyday. Everyday ethics involve what the media phenomenologist Tim Markham has called 'digital ethical inhabitation' (2020: 43), whereby 'Ethics emerge through . . . one's mundane navigation of digital environments' (Markham, 2020: 140). Thus, treating ethics as an ongoingly emergent facet of the everyday experiential world, rather than as principles that automated technologies can manifest, seems crucial if we are to understand how ethics and trust might configure in possible future automated systems, or worlds.

In this chapter, I first outline a design anthropological theory of everyday trust and ethics, supported by existing design ethnographic studies in possible transport mobility futures. I describe how trust is differently conceptualised in selected techno-solutionist public domain industry and policy renderings, and how this aligns with trust research in HCI research and innovation. Such dependencies fuel and fund a self-perpetuating research paradigm which does not account for the generative qualities of the everyday. Therefore, I propose an interdisciplinary agenda, which accounts for everyday trust as a realistic and plausible prism through which to comprehend how possible futures could play out.

Trust and ethics through design anthropology

Design anthropologists understand the everyday world as ongoingly emergent, a site of contingent circumstances where people develop routines and improvise as they go about living (Smith and Otto, 2016; Pink, 2021). This understanding concurs with phenomenological anthropologies of ethics, as outlined by Cheryl Mattingly and Jason Throop, whereby the human condition 'is at once excessive, uncertain, and emergent' (Mattingly and Throop, 2018: 482) and ethics are embedded in, and emergent from the everyday. Everyday ethics are thus concerned with what it is socially and culturally right to do, but rather than being fixed, they are relational and shift in particular circumstances, while they also, like social relations and culture, change over time. Ethics are thus underdetermined (Mattingly and Throop, 2018: 486), contingent, improvised to suit situations (and the limitations of these situations) as they unfold (Pels, 2000; Pink, 2017). Such ethics are sedimented in embodied ways of being and knowing that are learned, sensed and incrementally evolve; they shape and are shaped by everyday actions and the decisions they represent.

Trust, while less vigorously discussed in anthropology, can similarly be understood phenomenologically as emerging within the everyday. Trust is an anticipatory concept, which should be treated as 'a *feeling*, or category of feeling, which describes [a particular kind of] anticipatory sensations' (Pink, 2021). The anthropologist

Alberto Corsin criticises approaches that treat trust as an interactional or trans-actional relationship between two entities – such as when a person is supposed to trust a technology. There, he points out, 'Trust emerges as an epiphenomenon of social knowledge: what people's relationships look like after the fact of cognitive re-appraisals' (Corsin, 2011: 178). My collaborative research into how and why people experience trust in relation to technologies and data within everyday life as lived (e.g. Pink et al., 2018, 2021) suggests that to trust involves 'a sensory experi-ence of feeling or disposition towards something' rather than an explicit cognitive decision made in relation to a specific technology (Pink et al., 2018). This work dialogues with philosophical renderings of trust as 'Confidence based on familiar-ity' (Frederickson, 2016: 59) to suggest that trust is relational and 'a sensation often achieved through the accomplishment of mundane everyday routines' (Pink, 2021).

My approach also concurs with situated contextual understandings of trust as dependent on the morality of the stakeholders involved with a technology, rather than in the technology itself (Simon and Rieder, 2021). To trust, therefore, is not a fixed or finished interaction between two things. Rather, it is an always unfinished feeling. Elsewhere I argue trust is an anticipatory feeling, which goes beyond how we sense and is 'concerned with the continually emergent moments through which we slip over the edge of the present into the future'; in this definition, trust relates to 'how we feel as we move forward, and how we feel about what might be going to happen next'. Here, trust is a 'feeling between what we know and what we think we know', and 'a way of imagining-in-the-body, or a sensuous mode of anticipation' (Pink, 2021).

Ethnographic research into trust in AD cars, undertaken as people use cars with limited and advanced AD features and simulated AD cars within their eve-ryday routines, reported in detail elsewhere (including Pink et al., 2020; Lindgren et al., 2020) has revealed that everyday trust is situated, contingent and shifting. In brief, ethnographic immersion in people's everyday journeys showed us how they invested trust in memories and objects, how trust in AD features grew and declined in intensity along different stretches of road during the same journey and how it varied between different participants. Here, trust emerged as an embodied, experiential mode of anticipating safety, rather than a rational decision based on an evaluation of the ethics embedded in the design of a vehicle. This work alerts us to attend to the biographical, sensory, affective, social and material circumstances in which trust is generated, how it emerges from and becomes part of everyday rou-tines and actions, and it is never complete, or determined, but always on the move.

Conceptualised design anthropologically, ethics and trust are both experienced in the complex, contingent and always unfinished circumstances of everyday life as lived and are moreover experientially inextricable from and conceptually relational to each other. Trusting in a particular situation involves, in part, feeling sufficiently confident of the ethical integrity surrounding that situation. If we wish to under-stand how the relationality between people, trust, ethics and automation config-ures, we need to encounter people as they experience and perform trust and ethics in everyday worlds. For trust and ethics to plausibly and realistically be considered part of our automated futures, the processes, practices and future imaginaries of technology design need to be equally aligned to the everyday sites in which they are generated.

Trust in automation as techno-solutionism

In this section, I discuss how trust is understood within industry, policy and research agendas that are shaped by a techno-solutionist innovation paradigm. This narrative seeks to connect trust in automation with human ethics. However, in doing so, it disconnects the everyday worlds where ethics and trust are emergent and experienced from those where automated technology design and development are conceived. To illustrate this, I draw on examples of future imaginaries for automated transport mobilities recently advanced in policy and industry publications.

Perspectives that assume that automated technologies will solve societal problems tend to predict that 'innovations' that are becoming technologically possible will have particular future impact. For example, the Australia & New Zealand Driverless Vehicle Initiative states that 'Driverless vehicles have the potential to provide significant road safety, economic, environmental and social benefits, including improved social inclusion'. It goes on to claim that 'This technology will make driving easier and safer, allow people to be more productive and offer greater mobility to a wider range of people than ever before, reduce emissions, and ease congestion' (ADVI, no date). However, in the public sphere, internationally, the question of how to gain these benefits remains. For instance, the 2020 American Automobile Association survey reported that 'only one in ten drivers (12%) would trust riding in a self-driving car' and that 'Even more Americans – 28% – don't know how they feel about the technology' (AAA, 2020). The AAA's director of Automotive Engineering and Industry Relations stated that 'Knowing how people truly feel about self-driving cars will help the industry to identify the steps needed to move consumers towards greater acceptance'. Online or phone surveys rarely reveal how people truly *feel*. However, the results of this survey do suggest that there is a lack of alignment between technology design, the ambitions of policy and industry and everyday life.

Self-driving or AD cars are often thrust into the limelight when it comes to public discussions of AI, ethics and trust and the ensuing discussions likewise often reveal this disconnect between everyday ethics and trust and how these feelings are articulated in techno-solutionist narratives. AD cars have long existed in the popular imagination (Kröger, 2016), have been extensively hyped (Stilgoe, 2019) and have been involved in high-profile road accidents (e.g. Ash, 2017) and sensationalised in media headlines such as 'Could your self-driving car choose to kill you?' (Science Focus, 2020) and 'Driverless cars: Who should die in a crash?' (BBC News, 2018). In such narratives, AD cars are cited to invoke questions of AI ethics and trust through the frequently cited 'trolley problem', which was originally conceptualised through the ethical dilemma faced by a trolley driver who experiences brake failure when going down a valley and has to choose between one or another accident scenario involving people in the trolley's path (see Ash, 2017). Likewise, an IBM article on 'Building Trust in AI' uses an iteration of the trolley problem to ask how 'we can trust that it [AI] reflects human values', inviting readers to:

> Suppose there's a bus coming toward a driver who has to swerve to avoid being hit and seriously injured; however, the car will hit a baby if it swerves left and an elderly person if it swerves right – what should the autonomous car do?
>
> *(IBM no date)*

The trolley problem is often invoked in relation to the question of if people will trust AD cars, and has been used to endorse the idea that if machines are designed to be ethical, then people will accept and trust them. Because the ethical questions posed by conundrums such as the trolley problem resonates globally (but one would imagine, not universally), they provide a good case for quantitative analysis. An MIT Moral Machine experiment, which sought to understand human ethics relating to AD collected '40 million decisions in ten languages from millions of people in 233 countries and territories' (Awad et al., 2018: 59) when posing a version of the trolley problem online (see Pink et al., 2021 for a critique of the MIT Moral Machine). Yet, in such renderings, ethics become crystallised in the moment of making a decision on an online platform, in the MIT case, or in the case of the IBM conundrum where the driver has to choose between themself, a baby or senior person, our readings of a situation abstracted from any of the circumstances of everyday life in which it would actually be experienced. That is, the ethics of these situations are suspended outside, rather than situated within, the everyday. Subsequently, the use of the trolley problem as a probe through which to understand what ethics a machine needs to demonstrate in order for people to trust it, equally extracts trust from the everyday. Because the trolley problem is constructed as an ethical *problem* – it seeks an ethical *solution*. In contrast, design anthropologically, everyday ethics do not take the form of solutions, rather, ethics are always constitutive of and emergent from the specificities and contingent circumstances of life.

This lack of alignment is precisely related to the disconnect between the research and industry agendas noted earlier and everyday life as lived. The trolley problem is both painful and spectacular to consider and thus has an affective dimension. However, it describes circumstances that most people are extremely unlikely to encounter in an everyday life situation either in the present or in the future. In this sense, it falls out of the scope of an everyday ethics on a second count: it is just not the kind of thing we make ethical decisions about as our everyday lives unfold, and it is simply not the question which will decide if someone will ride in, and trust an AD car.

While AI tends to be discussed in relation to more spectacular life and death scenarios, such as AD car accidents, ADM is also implicated in techno-solutionist narratives within the transport mobilities field. For instance, a 2019 World Economic Forum (Ketter, 2019) report invites readers to 'imagine' a policy initiative whereby electric vehicle (EV) chargers were all installed with 'the potential for an external control, but that EV owners were put in charge of the decision to exert that control'. EVs are seen by many – including policy and industry organisations – as an environmentally sustainable solution which will lead to the reduction of carbon emissions from fossil fuel derived energy. However, the automotive and energy industries are faced with the problem of how to ensure that people will charge EVs in a way that optimises efficiency and distributes demand for energy without putting pressure on the energy grid. The WEF proposal is a solution to this problem. It asks readers to imagine a future where:

> a machine learning assistant, controlled by you and with the capacity to interact with real-time electricity markets, learns your preferences and offers you the flexibility to make choices based on your immediate preferences for cost savings, convenience or range security.

It suggests that letting machine learning assistants 'create the kind of flexibility and demand response that will help grid operators cope with increasing demand and supply volatility' both supports a sustainable energy transition with reference to EVs and 'extends to the integration of households that put energy they produce back into the grid' (Ketter, 2019). The WEF vision, presented as a clear message that future automation is the solution, on the surface appears to account for people and their need to be in control of how their decision-making is managed. Yet this vision involves a limited assumption regarding that future EV users' preferences will revolve around three particular choices: 'cost savings, convenience or range security'. It does not account for the everyday priorities, shaped by ethics – for instance, the ethics of care for others – that inform people's real choices above cost, convenience and range. It also fails to consider the contingencies of the everyday and that people often have long lives which are lived out in often uncertain, complex and changing circumstances and where their priorities are not rationally defined through issues of cost, convenience and range. Rather, as design anthropological studies have revealed, people's concerns are more likely to be social, ethical and nuanced. Everyday decisions are enacted in the midst of contingent situations which, in realistic and plausible futures, are likely to complicate such charging preferences.

The emergence of trust as ethics in the academic innovation agenda

Within the techno-solutionist narrative outlined earlier, trust is situated as something that *needs* to be generated so that people will (correctly) use automated technologies primed to solve societal problems. Subsequently, there is a demand for research and innovation that claims to design technologies that people will trust. This demand often underpins the economy of research funding in engineering and technology fields.

HCI research stretches across academic and industry research labs and funding sources. HCI research into human trust in technology has traditionally been informed by variations of the technology acceptance model (TAM) (Hombaek and Hertzum, 2017; Koul and Eydgahi, 2018), based on psychological theories of trust and acceptance, and quantitative methods. For instance, an 'analysis of 258 empirical HCI research articles on trust in automation and AVs [autonomous vehicles]' published between 1991 and 2019 led by Kaspar Raats concluded that 'research methodologies used for studying trust in automation and AVs, tend to rely on quantitative laboratory research and questionnaire based trust assessment' and that 'HCI trust research is conventionally undertaken according to a logic that trust evolves between a person and a machine in an interaction situation' (Raats et al., 2020). Thus, isolating trust as occurring between just two entities in controlled conditions, rather than situating trust as it occurs in everyday life.

As technologies are increasingly automated, trust has become incorporated into proposals to generate user trust in technologies through automated systems. For instance, blockchain technologies could be used to automate trust within automated EV charging solutions, such as that outlined in the WEF proposal discussed earlier, by guaranteeing validated trading partners. The blockchain validation of the trading partners would ensure that users could trust flexible charging systems (MIT Technology Review, 2019). An example of this is an EV 'charging scheduler application' which uses blockchain for transparency and verifiability and to create flexible charging times (Pajic et al., 2018: 263–4). Integrating EV charging into new energy systems also creates problems for which technical solutions are suggested. For example, Gorenflo et al. (2019) discuss how energy can be traded within new centralised energy systems, involving distributed generation, EVs and storage technology. However, they identify issues relating to trust between property owners and the charging service provider and between customers and the charging service provider. For their solution to work, they need customers to trust the charging service provider to 'record their charging events and bill them properly, as well as to record wallet top-ups' (Gorenflo et al., 2019: 161). They also need the property owners to trust the service provider to 'record charging events and reimburse them properly' (2019: 161). They propose using blockchain as a 'trusted central party' to make trusted 'direct peer-to-peer energy trading' within such systems possible (2019: 160), since it records and makes visible all the transactions to all parties (2019: 162). In both these proposals trust and flexibility are seen as technology design solutions, that mediate trust between humans, and which will be accepted and adopted by rational or self-interested consumers.

These shifts towards making AI and ADM trustworthy are reflected in some recent HCI research. In a 2021 issue of the *International Journal of Human-Computer Studies*, Donghee Shin (2021) suggests a new TAM is needed to account for users' unfamiliarity with AI. Shin focuses on three concepts: fairness, accountability, transparency and explainability (FATE) (Shin, 2021: 1) to account for trust. They propose that

> users evaluate explanations based on their existing knowledge and beliefs, and partly based on their understanding of the algorithms. . . . When such

explanations are reasonable and understandable, users begin to accept FAT [fairness, accountability and transparency] and trust the AI system.

(Shin, 2021: 2)

From the perspective of this disciplinary focus: 'The causal implications of trust and algorithmic explainability provide important directions for academia and practice' (Shin, 2021: 2), but how FATE is processed by 'users' or impacts trust has not been determined. Thus, '[t]he established relations between FATE and trust will be a stepping stone to further explore the role of FATE in AI design' (Shin, 2021: 7).

Critical approaches in design and HCI research contest the causality in such paradigms. Elisa Giaccardi and Johan Redström's argument, from a more-than-human design perspective, alerts us to the important point that 'ideas of full "transparency", "explainability", and "trust-worthiness" of how networked computational things operate and relate assumes a centrality of human agency and intentionality – only humans make and use things and can therefore control them' (Giaccardi and Redström, 2020: 41). Instead, acknowledging the active roles that things play in the ways relations configure, they argue that

for some transparency, explainability, and trustworthiness to be in place, things will need to be designed so they can continue to generate affordance and value in ways that can be negotiated as appropriate under circumstances of use that are always meant to change.

(Giaccardi and Redström, 2020: 41)

Similarly, in the field of 'entanglement HCI', Frauenberger (2019) has argued for an approach that would 'decentre the human as the sole source of activity and . . . elevate the role of the non-human world from a passive backdrop to human activity, to active contributors to relational action as it unfolds' (Frauenberger, 2019: 21).

In common with design anthropology, these approaches emphasise a theory of the everyday as a site of ongoing emergence, where trust is situated not in finished products but in the unfinished and changing circumstances where technologies unfold. The implication is that ADM and AI need to be designed to participate in and to be responsive to the sites of the everyday where ethics and trust come about.

Automation and the temporalities of innovation

In the techno-solutionist narratives outlined earlier, human trust is a causal outcome of a technology design process that successfully invests human ethics in automated technologies. Following this logic, human trust subsequently leads to the predicted benefits of technological innovation in ADM and AI supported technologies. This vision of innovation has a particular temporality which underpins research agendas and funding and, I argue, creates a divide between well-intentioned computer and engineering sciences working for 'social good' and the ethics of the everyday.

The anthropologists Tim Ingold and Elizabeth Hallam (Ingold and Hallam, 2007, 2) have defined innovation as an 'after-the-event' concept, where it refers to a finished product, arguing instead for an approach that accounts for the ongoingness of things and processes, whereby they are never complete. Science and technology studies scholars have similarly called for a focus on 'what happens *after* innovation' (Russell and Vinsel, 2016). These critical approaches have shown us that we need to account for innovation narratives as retrospective devices that seek to close off processes as finished, in wait for the next innovation. However, narratives about emerging technologies that define innovation predictively – such as those discussed earlier relating to the future of AD cars and EV charging – appear to indicate a different temporality and require a new academic response: they are 'before-the-event' innovation narratives which leave no space for uncertainty that emerging technologies will have future impact because solutions that will guarantee their acceptance by society are proclaimed to be in sight.

Before-the-event innovation narratives pursue innovation to the only conclusions that the logics of the solutionist paradigm can offer – they envisage a future that could be possible if only people will trust and appropriately use the finished technologies made by engineers and rolled out by government and industry. The next step in this narrative is to design these technologies so that the causal relationships between ethics and trust, described in the previous section, will come about. For example Data61, part of Australia's national science agency – CSIRO – advises that 'Artificial Intelligence (AI) has the potential to increase productivity, create new industries and provide more inclusive services' but that 'For Australia to realise these benefits however, it will be important for citizens to have trust in the AI applications developed by businesses, governments and academia'. Data61's website suggests that 'One way to achieve this is to align the design and application of AI with ethical and inclusive values' (Data61, no date). To ensure that this future is arrived at, funds are invested in engineering and computer science research and development projects which aim to design ethical AI and ADM that people will trust.

Within this before-the-event or predictive innovation agenda, as Birhane has expressed it: 'challenges that refuse such a problem/solution formulation, or those with no clear "solutions", or approaches that primarily offer critical analysis are systematically discarded and perceived as out of the scope of these fields' (2021: 1). Put another way, there is an erroneous research emphasis on making automated technologies that are trustworthy because they embody human ethics, enable human rights and are transparent *instead of* engaging with questions about how to design automated systems that will become part of ethical, just and meaningful future environments where human rights are respected. Take, for example, the IBM article on 'Trust in AI' cited earlier, for which they 'spoke to 30 AI scientists and leading thinkers'. The article reports that '[t]hey told us that building trust in AI will require a significant effort to instill in it a sense of morality, operate in full transparency and provide education about the opportunities it will create for business and consumers'. The gist of this vision is that trust in AI can be generated through the design of transparent and ethical AI and then educating people to use it properly as they learn to live with it. It is also, from a methods perspective, unsettling to read of a research approach that 'spoke to' rather than listened to people, and only spoke to scientists and leading thinkers in the AI field, rather than to the people who would be educated to live with the AI.

The technological possibilities of AI and ADM, when seen as before-the-event innovations, create *anticipatory infrastructures* (Pink et al., 2022). Put another way, they not only provide the algorithms, capabilities and systems through which automation happens but also deliver the architectures through which 'better' futures facilitated by techno-solutionist pathways can be imagined. The belief that ethics might form part of these prescriptive frameworks should not be surprising to social scientists: anticipatory ethics are already integral to our lives as academics in the form of university ethical approval committees. Yet as anthropologists of ethics have revealed, written codes of ethics and the regulatory procedures attached to them are unaligned with the everyday ethics that inform how research is undertaken (reviewed in Pink, 2017). Likewise, the ethics that are associated with explainable and transparent AI are abstracted from everyday life. Like university ethical approval procedures, when ethics are designed into before-the-event innovations in the form of ethical AI or ADM, they are anticipatory risk mitigation strategies, which would mitigate the risk of humans not trusting automated technologies, or believing they cannot be trusted.

Anthropology and design offer us clear messages about ethics. From anthropology, Mattingly and Throop ask what it means 'to portray the human as characterised by potentiality or possibility rather than actuality?' and 'to claim that there is an excessiveness to the ethical demand such that it cannot be reduced to following prescriptive norms or rules?' (2018: 486). Here, ethics are emergent from the circumstances of life, and at the same time they constitute the values through which quotidian practices, experiences and imaginaries play out. Therefore, the ethics of how people will use AD cars will pivot on their feelings and priorities in the moment, rather than a fixed decision-making code. This is a stark contrast to the extractivist approach to ethics which gathers them for moments of the everyday in the form of questionnaire responses, or in the example of the MIT 'Moral Machine' by inviting people to respond to an unlikely dilemma abstracted from the everyday on an online platform. From design, locating ethics in everyday worlds where people and technologies evolve together, Giaccardi and Redström emphasise,

> the point of gravity for an ethical uptake of design is not to be located in the delegated functionality of the thing to be used (i.e., the what of intended use) but in the co-performance of people and things (i.e., the how of the relation).
>
> *(Giaccardi and Redström, 2020: 41)*

Posing Mattingly and Throop's questions (above) against causal logics that assert that quantified ethical ADM and AI will invoke human trust, reveals how these logics are deeply disconnected from the living world of everyday ethics and from the sites where trust is felt rather than rationally evaluated.

Interdisciplinary ethics and trust?

In this chapter, an interrogation of the concept of trust has created a prism through which to view how the logics of everyday ethics both live with and complicate those of techno-solutionism. Can trust subsequently be re-focused as a category for new interdisciplinary dialogue and collaboration between the sites of the everyday and the research and innovation projects of computer and engineering sciences?

The question of trust in automation is problematic from a number of perspectives. First, the fact that trust has been a key conundrum in HCI research for so long alerts us to the possibility that it is a problem that has so far evaded a solution from the perspective of its own disciplinary paradigms. Second, trust in automation is an awkward conceptualisation in that anthropologically it doesn't add up with either the theoretical or empirical evidence. At worst, the idea that people might trust in ethical automated technologies as intended could have a dystopian *finale*, whereby if its logic were to play out, then we would ultimately live in an implausible world of quantifiable interactions and values which were measures of trust. More realistically, the assumption that trust is rational sentiment which can be induced in people by ethical technologies (which appear to exhibit extracted ethics) simply leads to a research and innovation agenda which burns time and money in pursuing the impossible quest of 'solving' or 'improving' situations which it has framed as 'problems'.

Design anthropology calls for a new conceptualisation of trust, as contingent, always incomplete and similarly meant to change as the 'things' that figure in Giaccardi and Redström's (2020) notion of the everyday. Elsewhere I have argued that 'Understanding trust as sensed means that if seeking to design for a future in which trust figures, we need to consider how to create the circumstances in which it can be felt' (Pink, 2021: 200). Bringing this together with the more-than-human perspective of Giaccardi and Redström and the awareness in HCI that Shin (2021) alerts us to, that not enough is yet known about how trust configures in relation to FATE, opens a window for interdisciplinary collaboration. This should ideally happen at this moment where the uncertainties that have been created by the emergence of new automated technologies and systems mean that the problems and solutions are still by no means as clear as they have appeared to be in the past.

This agenda requires new openness to interdisciplinary collaboration, theories of the relationship between everyday and technological ethics and trust, flows between anthropology, design and technology development located in the sites of the everyday where such processes and actors are co-situated. It requires a research and innovation agenda which is inspired by the plausibility of possibilities rather than the promise of solutions and that treats the social and engineering sciences as equally pivotal in our futures. Automation is a central question of our times, for all disciplines and multiple stakeholders, as are issues of ethics and trust. It as such offers an opportunity to shift the agenda through sustained interdisciplinary attention in these spaces of concern.

Acknowledgements

The theoretical research discussed in this chapter was conducted as part of Sarah Pink's research in the ARC Centre of Excellence for Automated Decision-Making and Society and funded by the Australian Government through the Australian Research Council. The research projects referred to in this chapter were undertaken collaboratively with colleagues at Halmstad University in Sweden and are referenced through our publications. Sarah Pink's contributions to those publications were undertaken as part of her roles in: the Human Experiences and expectations of Autonomous Driving (HEAD) project funded by Swedish Innovation Agency VINNOVA (2016–2018); and Co-designing future smart urban mobility services – A Human Approach (AHA) and Design Ethnographic Living Labs for Future Urban Mobility – A Human Approach (AHA II), funded through Drive Sweden by the Swedish Innovation Agency Vinnova, the Swedish Research Council Formas and the Swedish Energy Agency (2018–2022).

References

AAA Newsroom (2020) Self-Driving Cars Stuck in Neutral on the Road to Acceptance. Available at: https://newsroom.aaa.com/2020/03/self-driving-cars-stuck-in-neutral-on-the-road-to-acceptance/

ADVI (n.d.) Driverless Car Benefits. Available at: https://advi.org.au/driverless-technology/driverless-car-benefits/

AlgorithmWatch (2019) Automating Society – Taking Stock of Automated Decision-Making in the EU. Available at: https://algorithmwatch.org/en/automating-society-2019/

AlgorithmWatch (2020) Automated Decision-Making Systems in the COVID-19 Pandemic: A European Perspective. Available at: https://algorithmwatch.org/en/automating-society-2020-covid19/

Ash J (2017) *Phase Media: Space, Time and the Politics of Smart Objects*. London: Bloomsbury.

Awad E, Dsouza S, Kim R, et al. (2018) The Moral Machine Experiment. *Nature* 563: 59–64. https://doi.org/10.1038/s41586-018-0637-6

BBC News (2018) Driverless Cars: Who Should Die in a Crash? Available at: www.bbc.com/news/technology-45991093

Birhane A (2021) Algorithmic Injustice: A Relational Ethics Approach. *Perspective* 2(2): 100205.

Corsin Jimenez A (2011) Trust in Anthropology. *Anthropological Theory* 11(2): 177–96.

Data61 (n.d.) Artificial Intelligence: Australia's Ethics Framework. Available at: https://data61.csiro.au/en/Our-Research/Our-Work/AI-Framework

Fredricksen M (2016) Divided Uncertainty: A Phenomenology of Trust, Risk and Confidence. In: Jagd S and Fuglsang L (eds) *Trust, Organisations and Social Interaction*. Edward Elgar Publishers.

Frauenberger C (2019) Entanglement HCI the Next Wave? *ACM Transactions on Computer-Human Interaction* 27(1): 2:1–2:27. https://doi.org/10.1145/3364998

Giaccardi E and Redström J (2020) Technology and More-than-Human Design. *Design Issues* 36(4): 33–44.

Gorenflo C, Golab L and Keshav S (2019) Mitigating Trust Issues in Electric Vehicle Charging Using a Blockchain. In: *Proceedings of the Tenth ACM International Conference on Future Energy Systems (e-Energy '19). Association for Computing Machinery*. New York, NY, 160–164. https://doi.org/10.1145/3307772.3328283

Hombaek K and Hertzum M (2017) Technology Acceptance and User Experience: A Review of the Experiential Component in HCI. *ACM Transactions Computer-Human Interaction* 24(5): 1–30. https://doi.org/10.1145/3127358

IBM (n.d.) Building Trust in AI. Available at: www.ibm.com/watson/advantage-reports/future-of-artificial-intelligence/building-trust-in-ai.html

Ingold T and Hallam E (2007) Creativity and Cultural Improvisation: An Introduction. In: Hallam E and Ingold T (eds) *Creativity and Cultural Improvisation*. Oxford: Berg, 1–24.

Ketter W (2019) How We Can Embrace the Electrical Vehicle Transition by Adopting Smart Charging. *World Economic Forum*. Available at: www.weforum.org/agenda/2019/05/how-charging-for-electricity-on-a-sliding-scale-could-power-the-electric-vehicle-transition/

Komendantskaya E, Arnaboldi L and Daggitt M (2021) Perfecting Self-driving Cars – Can It Be Done? *The Conversation*. Available at: https://theconversation.com/perfecting-self-driving-cars-can-it-be-done-157483?utm_source=twitter&utm_medium=bylinetwitterbutton

Koul S and Eydgahi A (2018) Utilizing Technology Acceptance Model (TAM) for Driverless Car Technology Adoption. *Journal of Technology Management & Innovation* 13(4): 37–46.

Kröger F (2016) Automated Driving in Its Social, Historical and Cultural Contexts. In: Maurer M, Gerdes J, Lenz B and Winner H (eds) *Autonomous Driving*. Berlin; Heidelberg: Springer.

Lindgren T, Fors V, Pink S and Osz K (2020) Anticipatory Experience in Everyday Autonomous Driving. *Personal and Ubiquitous Computing* 24: 747–62.

Markham T (2020) *Digital Life*. Cambridge: Polity.

Mattingly C and Throop J (2018) The Anthropology of Ethics and Morality. *Annual Review of Anthropology* 47(1): 475–92.

MIT Technology Review (2019) Available at: https://www.technologyreview.com/10-breakthrough-technologies/2019/

Morozov E (2013) *To Save Everything, Click Here: Technology, Solutionism, and the Urge to Fix Problems That Don't Exist*. London: Penguin Books.

Pajic J, Rivera J, Zhang K and Jacobsen H (2018) EVA: Fair and Auditable Electric Vehicle Charging Service Using Blockchain. *e-Energy '19: Proceedings of the Tenth ACM International Conference on Future Energy System*, 262–5. https://doi.org/10.1145/3210284.3219776

Pels P (2000) The Trickster's Dilemma: Ethics and the Technologies of the Anthropological Self. In: Strathern M (ed) *Audit Cultures: Anthropological Studies in Accountability*. London: Routledge.

Pink S (2017) Ethics in a Changing World: Embracing Uncertainty, Understanding Futures, and Making Responsible Interventions. In: Pink S, Fors V and O'Dell T (eds) *Working in the Between: Theoretical Scholarship and Applied Practice*. Oxford: Berghahn.

Pink S (2021) Sensuous Futures: Re-thinking the Concept of Trust in Design Anthropology. *Senses & Society* 6(2): 193-202.

Pink S, Dahlgren K, Strengers Y and Nicholls L (2022) Anticipatory Infrastructures, Emerging Technologies and Visions of Energy Futures. In: Valkonen J, Kinnunen V, Huilaja H and Loikkanen T (eds) *Infrastructural Being: A Naturecultural Approach*. Basingstoke, UK: Palgrave Macmillan.

Pink S, Lanzeni D and Horst H (2018) Data Anxieties: Finding Trust and Hope in Digital Mess. *Big Data and Society* 5(1) https://doi.org/10.1177/2053951718756685

Pink S, Osz K, Raats K, Lindgren T and Fors V (2020) Design Anthropology for Emerging Technologies: Trust and Sharing in Autonomous Driving Futures. *Design Studies* 69. https://doi.org/10.1016/j.destud.2020.04.002

Pink S, Raats K, Lindgren T, Osz K and Fors V (2021) An Interventional Design Anthropology of Emerging Technologies. In: Bruun MH, Wahlberg A, Kristensen DB, Douglas-Jones R, Hasse C, Høyer K and Winthereik BR (eds) *The Handbook for the Anthropology of Technology*. Palgrave Macmillan.

Raats K, Fors V and Pink S (2020) Trusting Autonomous Vehicles: An Interdisciplinary Approach. *Transportation Research Interdisciplinary Perspectives* 7: 100201.

Russell A and Vinsel L (2016) Hail the Maintainers. *Aeon*. Available at: https://aeon.co/essays/innovation-is-overvalued-maintenance-often-matters-more

Science Focus (2020) Could Your Self-driving Car Choose to Kill You? Available at: www.sciencefocus.com/future-technology/could-your-self-driving-car-choose-to-kill-you/

Shin D (2021) The Effects of Explainability and Causability on Perception, Trust, and Acceptance: Implications for Explainable AI. *International Journal of Human-Computer Studies* 146.

Simon J and Rieder G (2021) Trusting the Corona-Warn-App? Contemplations on Trust and Trustworthiness at the Intersection of Technology, Politics and Public Debate. *European Journal of Communication*. https://doi.org/10.1177/02673231211028377

Smith RC and Otto T (2016) Cultures of the Future: Emergence and Intervention in Design Anthropology. In: Smith RC, Vangkilde KT, Kjærsgaard MG, Otto T, Halse J and Binder T (eds) *Design Anthropological Futures*. London: Bloomsbury Academic, 19–36.

Stilgoe J (2019) Self-driving Cars Will Take a While to Get Right. *Nature Machine Intelligence* 1: 202–3. https://doi.org/10.1038/s42256-019-0046-z

Tomašev N, Cornebise J, Hutter F, et al. (2020) AI for Social Good: Unlocking the Opportunity for Positive Impact. *Nature Communication* 11: 2468. https://doi.org/10.1038/s41467-020-15871-z

3

THE QUANTIFIED PANDEMIC

Digitised surveillance, containment and care in response to the COVID-19 crisis

Deborah Lupton

Introduction

The COVID-19 outbreak first emerged in the final weeks of 2019 and began to rapidly spread globally from early 2020. By 11 March 2021, COVID-19 had been officially declared a pandemic by the World Health Organization (WHO). A year later, WHO's dashboard showed that over 116 million cases of infection with the novel coronavirus (SARS-CoV-2) that causes the disease COVID-19 had been confirmed globally, with more than 2.5 million deaths (World Health Organization, 2021). This dashboard, displayed on WHO's website and updated daily, is only one example of the online displays of COVID-related data that have played a major role throughout the pandemic in informing authorities and publics of the impact of the novel coronavirus across the world. During the first year of the pandemic, a wide array of other digital technologies were repurposed, invented or proposed to monitor, prevent or treat COVID infections. These technologies include apps used to monitor people in quarantine and self-isolation, contact tracing apps, surveillance drones, digitised temperature checking devices, apps for delivering COVID test results, software for identifying 'at risk' patients and for selecting recipients of vaccines and digital vaccine 'passport' apps, as well as automated symptom checker apps, platforms and chatbots designed to help people determine whether they were infected with the novel coronavirus or needed to seek medical attention.

In this chapter, I present a sociocultural analysis of how such technologies were deployed or anticipated in the 12 months following WHO's pandemic declaration. In doing so, I build on scholarship in critical public health, technocultures and critical data studies to discuss these technologies' social and political contexts and effects. The discussion not only directs a particular focus on automated decision-making (ADM) systems but also includes related technologies. It is important to emphasise that ADM technologies operate within non-ADM systems: not least because there

DOI: 10.4324/9781003170884-5

is continuing lack of clarity of how to define and distinguish ADM from automated or partly automated systems or from AI (AlgorithmWatch, 2020; Elish and Watkins, 2020). It is not always possible to peer inside the 'black box' of emerging technologies to determine to what extent they use automated systems. Many developers of emerging technologies claim they are automated when in fact extensive human intervention is required for their operation (Elish and Watkins, 2020). Furthermore, there are manifold porosities and exchanges between ADM and other digitised devices and software infrastructures applied to the COVID crisis, including the discourses, logics, imaginaries and material practices involved and their impact on modes of governance and populations. For these reasons, rather than seek to define ADM or single out ADM technologies for attention in this chapter, I provide an expansive overview.

The field of critical public health studies offers a broad perspective that contextualises the emergence of use of digital technologies for COVID surveillance, control and care within analyses of public health attempts to combat disease at the population level (Lupton, 1994, 1995, 2021). It focuses, in particular, on the social and political dimensions of public health initiatives: including digital health technologies (Lupton, 2014, 2017b). Critical public health studies scholarship highlights the focus on personal responsibility for maintaining and protecting good health, an associated tendency in public health discourses towards blame and moralising in relation to becoming ill or being designed as 'at risk' from disease, and the stigmatisation and marginalisation of disadvantaged social groups that have been integral to discourses and practices of infectious disease control for centuries (Lupton, 1994, 1995; Mack, 1991; Bashford, 2016).

The key concepts of digitisation (Lupton, 2017a), datafication (van Dijck, 2014) and dataveillance (Raley, 2013) are drawn from critical data studies. Digitisation refers to the use of digital technologies to render aspects of living and non-living phenomena into digital formats. In the case of the COVID-19 crisis, aspects of people's health status, embodiment, movements in space and place and everyday practices are digitised. These processes of digitisation generate quantified data and, therefore, result in the datafication of human bodies and health states. Dataveillance is a term used to describe the ways that strategies for generating and processing personal information (these days often derived from the digital traces left when people go online, use mobile devices and apps or move around in sensor-embedded spaces) are used to monitor and watch people for purposes such as health or security surveillance.

The technocultures literature identifies the logics (Mol, 2008) and sociotechnical imaginaries (Jasanoff, 2015) that are central to the ways that digital devices and software for COVID management have been invented, promoted and proposed. Together, these logics and imaginaries operate to present a techno-utopian vision involving promissory narratives in which novel technologies are presented as the way of the future, offering greater time and budgetary efficiencies and better accuracy compared with previous approaches (Lupton, 2014, 2017b). Studies of people's lived experiences of digital health have pointed to the messiness of incorporating these technologies into everyday life, demonstrating how these promises and imaginaries are often confounded, disrupted or resisted when they enter the 'real world' of everyday life (Lupton, 2014, 2017b). Bringing these perspectives and concepts together, a central tenet of my argument in this chapter is that digital technologies directed at COVID control, management and care are always already sociomaterial assemblages of humans and non-human agents (Lupton, 2019). In the time of a typically once-in-a-century crisis such as the COVID-19 pandemic, these assemblages combine compelling discourses, logics, imaginaries and affects with digital devices and software, fleshy human bodies, the novel coronavirus, vaccines, place and space and a plethora of other non-human agents. The development, promotion and enactments of digital technologies such as ADM and related software in response to COVID cannot be understood without acknowledging the human and other more-than-human (and more-than-digital) elements of these assemblages.

Pre-digital infectious disease surveillance, control and care

Historical and sociological analyses of public health demonstrate that a tension between individual rights and the public health good has been evident for centuries in public health control and management of infectious disease. The institution of public health is focused on the management of populations so as to best contain the spread of communicable diseases or prevent against people from developing non-communicable diseases such as lung cancer, heart disease or diabetes (in recent times, often referred to as 'lifestyle' diseases) (Lupton, 1994, 1995). Traditional quarantine measures, involving the physical isolation of people deemed to be infected with a contagious illness or those who have had close contact with infected people, have been employed since the 14th century for disease control (Bashford, 2016; Mack, 1991). In more recent times, neoliberal approaches to healthcare and public health have tended to focus attention on individual responsibility for achieving and maintaining good health, including engaging in preventive health behaviours (Lupton, 1994, 1995). Simultaneously, however, public health acts in many countries have allowed for international border closures, travel restrictions and the enforced isolation or even imposing significant fines on or detention of people considered to pose a serious threat to others by virtue of their infectious status (Martin, 2006).

Well before the digitisation of public health and medicine, datafication and dataveillance had been central strategies for the containment of infectious diseases for centuries. Particularly, since the rise of epidemiological surveillance of populations in the 18th century as a biopolitical initiative (Foucault, 1984), these strategies have been employed to assist in collecting, archiving, processing and displaying health-related data such as cases of disease and deaths (Armstrong, 1995) as well as for infectious disease management efforts such as contact tracing during epidemics (Kahn, 2020). Critical scholars have drawn on Foucault's work on biopolitics and governmentality (Foucault, 1984, 1991) in analysing these endeavours, including the role of public health surveillance and the promotion of personal responsibility for good health and risk avoidance in governing populations. They have shown how public health measures for dealing with infectious and other diseases rest on identifying risk behaviours and risky places as well as the social groups who are most vulnerable to contracting and spreading a disease (Lupton, 1994, 1995, 2013; Armstrong, 1995).

These strategies of documentation and quantification have operated in the interests of taming uncertainty (Hacking, 1990) and seeking to exert control over a situation in which risk appears to be rampant (Lupton, 1994, 1995, 2013). Cultural risk theory has demonstrated how concepts of Otherness are symbolically and materially maintained and reinforced during health emergencies and crises (Lupton, 1994, 1995, 2013; Douglas, 1992). Time and again, medical historians and legal scholars have drawn attention to how public health surveillance, regulations and restrictions have often targeted stigmatised outgroups such as immigrants, people of colour and the poor as posing a threat to more privileged groups and requiring greater control and closer surveillance (Bashford, 2016; Mack, 1991; Martin, 2006).

Since the emergence of personal and mobile computing, the internet and Wi-Fi, practices of monitoring and measuring human health and disease patterns have increasingly become datafied and digitised. Tens of thousands of mobile apps have been released for mobile devices and wearable devices such as smartwatches that provide opportunities for users to engage in self-diagnosis, track their bodily functions and activities and share their metrics with others online. Telehealth platforms are available in some countries for medical practitioners to engage with patients and epidemiology is often supported by digital tools for gathering data about disease patterns (Lupton, 2017b). Some of these digitised health systems are promoted as offering ADM capabilities. However, as detailed investigations of the extent to which and how such software is actually deployed 'in the wild' (AlgorithmWatch, 2020) together with ethnographic analyses of practices of use (Elish and Watkins, 2020) have shown, such systems are typically not as 'automated' as they claim. They require continual and often invisible support and repair work from people for their successful operation.

Digitised COVID-19

As the COVID crisis unfolded over 2020, quantification and modelling to generate predictions of how COVID might spread if unchecked were vitally important to public health knowledges and government policies (Rhodes et al., 2020; Milan, 2020; Milan and Treré, 2020). This is particularly the case with presenting online dashboards and announcements of regularly updated metrics of COVID cases and deaths, such as the WHO COVID dashboard mentioned earlier, which have helped health authorities, governments and publics to understand the pattern of COVID spread. These metrics have contributed further to efforts to quantify and predict the futures of COVID spread and subsequent policy development concerning measures such as restrictions, border closures and lockdowns.

From the beginning of the pandemic, graphs of rises and falls in active cases and the number of deaths, recovered cases and so on have been constantly broadcast in the news media; predictions based on modelling have led public health policy; and specific aspirational metrics have been set as aims as part of 'road maps' for exiting lockdowns and other restrictions. In many countries, health authorities and politicians have featured in press conferences (daily at some points in the pandemic), providing the figures for a region or country for that day as a way of demonstrating to the public both that they understand how the outbreak is spreading or slowing and that they are taking action to 'flatten the curve' (a metaphor that itself relies on understandings of graphs and epidemiological patterns of disease spread). As a consequence, as Rhodes et al. (2020: 255, original emphasis) have observed: 'We have come to know COVID-19 infection control as a *calculation*'.

Critical studies of quantification as a mode of monitoring, governing and managing populations (Hacking, 1990) have shown that despite its claims to objectivity and verisimilitude, metricisation is inherently a social and political strategy. Given the oft-held assumption that digital modes of datafication are far more accurate, scientific and inclusive than human-led and analogue forms of quantification and record-keeping, the introduction of digital tools for datafying and visualising elements of human bodies such as disease and death rates has only contributed to these claims (Wernimont, 2019; Lupton, 2016, 2017b). As with all big datasets, there is a diverse array of factors that can cause inaccuracies in datafied tools such as WHO's COVID tracking dashboard. These metrics are themselves dependent on factors such as COVID testing rates and accuracy, the ways that COVID deaths are identified and categorised, and efficient reporting of cases and mortality: all of these strategies have become highly politicised. For example, former US President Trump was notorious for making public pronouncements about wanting to slow down COVID testing rates in the USA so that the country's case statistics would not appear as dire (ABC News Online, 2020). Authoritarian and corrupt governments have been called to account for continual under-testing and under-reporting of COVID cases and mortality rates, while other countries simply do not have the resources to offer comprehensive COVID testing (Winter, 2020; Milan and Treré, 2020).

Digitisation, datafication and dataveillance have played a central part in monitoring and disciplining people placed in quarantine as well as physical distancing and self-isolation measures introduced to control the spread of the novel coronavirus. Many contact tracing apps have been released globally that involve recording users' movements and locations and have been developed especially for COVID surveillance efforts. The extent to which these apps are automated varies greatly, as does their invasiveness of privacy. Some of these apps involve ADM systems and highly detailed centralised collection of geolocation data, while others just record proximity information, using Bluetooth 'handshake' technologies that record details of close contact with other people using the app (AlgorithmWatch, 2020; Kahn, 2020). Regardless of the extent of automation used, such apps have routinely been portrayed by government authorities as integral to controlling the spread of the virus. For example, when Australia's COVIDSafe contact tracing app was released in April 2020, the Australian Prime Minister, Scott Morrison, claimed that downloading it was Australians' 'way out' of the pandemic. (It proved quite useless and soon was no longer mentioned by Morrison.)

In some locations, digital technologies and digital data analytics have been taken up as ways of tracking people's location and movements to ensure that they adhere to self-isolation restrictions for the length of the quarantine period. Authoritarian governments, in particular, have introduced mandatory dataveillance systems to combat COVID, with little interest in how such systems may flout human rights (AlgorithmWatch, 2020). In China, people were prevented from leaving their homes if they had been identified as infected with COVID by an automated rating system on an Alipay or WeChat phone app that coded them 'red' and, therefore, as requiring to go into quarantine (Zhou, 2020). In some Chinese cities, local government authorities have brought in automated monitoring measures using facial recognition data and smartphone data tracking combined with information derived by requesting people to enter details about their health and travel history into online forms when visiting public places (Goh, 2020).

It was not only Chinese authorities who introduced invasive digitised forms of identifying infection risk and enforcing isolation. But South Korea, Israel, Hong Kong and Taiwan also deployed similar tools for surveillance of the movements of people who had been diagnosed with the coronavirus, their contacts and other people ordered to be in quarantine, including those newly arrived from overseas. South Korea used an app that informed potential contacts of an infected person, whereas Hong Kong issued incoming travellers with tracking bracelets that would automatically monitor their geolocation (Calvo et al., 2020). Russia introduced automated facial recognition systems early in the pandemic, used by police to identify and fine people who have broken regulations about self-isolation and physical distancing (AlgorithmWatch, 2020). This country also compels people who have tested positive for COVID to install a geolocation tracking app, titled Social Monitoring, that is automated to send notifications every two hours for the user to upload a selfie to prove they are at home, even throughout the night. People who fail to comply are issued with fines. A similar app was used in Poland (Gershgorn, 2021). Nationally mandated location tracking apps to monitor the movements of citizens were also introduced in Qatar, India, Russia and Poland early in the pandemic and were still in use in early 2021. People in some of these countries were denied entry to businesses, government offices or public transport unless they could demonstrate that they checked in using the app (Gershgorn, 2021).

ADM systems and other digital technologies for COVID are not limited to dataveillance initiatives. A multitude of digital technologies have been introduced for offering diagnosis and medical care, building on a diverse range of digital health technologies that had been in development or available for use well before the pandemic. These include digitised temperature checking devices, apps for delivering COVID test results, software for identifying 'at risk' patients and for selecting recipients of vaccines, and digital vaccine 'passport' apps, as well as automated symptom checker apps, platforms and chatbots designed to help people determine whether they were infected with the novel coronavirus or needed to seek medical attention. Some of these systems use very simple automation that does not involve a high level of ADM. The software is programmed to send notifications or to 'decide' whether someone has a fever above a defined 'risk' level or is eligible for a vaccine, for example. Others rely on more complex software that is programmed to respond to queries and, therefore, makes 'decisions' about the correct response. For example, the US Centers for Disease Control and Prevention provided an online assessment chatbot for people, Coronavirus Self-Checker bot, that drew on Microsoft's Healthcare Bot service. It asked questions about symptoms designed to help users decide whether they needed to seek medical advice but did not provide a diagnosis. Several tech companies marketed 'fever detection' systems that involve different decision-making software operating together: in some cases, combining thermal sensors with facial recognition and algorithmic sorting systems. The US-based Feevr company, for example, claimed on its website that 'Feevr is a quick and effective AI-based system for screening and detecting individuals with elevated temperature in a crowd' (feevr.tech, 2021).

By March 2020, a 'pandemic drone' was in the process of development by the University of South Australia. It was fitted with sensors and a machine vision system with advanced pattern recognition software. The developers claimed that the drone would be able to monitor temperature, heart and respiratory rates within 5–10 metres as well as people sneezing or coughing in public places (Olle and AAP, 2020). Later shelved due to privacy concerns, the developers found a new use for this technology, employing the sensors in COVID monitoring stations at Alabama State University in the USA (Spence, 2021).

Digital surveillance, function creep and algorithmic injustices

Underlying the apparent convenience offered by digital technologies for COVID surveillance, control and care are significant failures. It is notable that several comprehensive reviews of digitised systems for COVID control, management and care, including those attempting to engage ADM, identified few successful strategies (AlgorithmWatch, 2020; Centre for Data Ethics and Innovation, 2021; White and Van Basshuysen, 2021). Offering diagnosis online is a high stakes intervention. While such tools can provide much-needed reassurance to people who are worried about whether they are infected and can take pressure off an overloaded healthcare system, other problems can result. Offering tools that devolve diagnosis to ADM without human intervention raises the risks of the software failing to accurately inform people of their risk, potentially wasting valuable health resources or alternatively, excluding people who need it from testing and further treatment. Similarly, given that many people infected with SARS-CoV-2 are asymptomatic and that many other illnesses cause a raised body temperature, while it has been widely used in public spaces such as healthcare sites, airports, workplaces and gyms, digital body temperature monitoring is more of a symbolic gesture of risk avoidance.

Beyond the questions of efficacy and accuracy, the uses of such technologies in response to the pandemic raise issues concerning how the logics, imaginaries and practices involved in the deployment of data-driven and predictive software can generate decisions that clash with human rights, human agency and personal data privacy. As is the case with traditional public health measures, the freedoms and autonomy of those deemed to be infected or at risk of infection are in tension with public health goals to control epidemics and pandemics. COVID dataveillance strategies are redolent of the measures that are used in the criminal justice system, where employing electronic monitoring technologies such as digital tracking bands has been deployed as a way of watching and containing offenders' movements once released from a custodial sentence (Graham and McIvor, 2017). In the Australian city of Adelaide early in the pandemic, two Chinese visitors identified as being infected with COVID were placed under voluntary home isolation, their movements monitored by the police using their smartphone metadata. It is notable that the police emphasised that this is the same dataveillance system used for tracking offenders in criminal investigations (Sutton, 2020).

The deployment of these technologies portends an ever-expanding reach into people's private lives and movements by health authorities and other government agencies that could continue well beyond the initial rationales of controlling the COVID crisis. Academic researchers and civil society groups have drawn attention to the issues with human rights and 'function creep' possibilities of these technologies, including the lack of protection of personal data, transparency about how authorities are currently using this information and plan to do so in the future and the inability of people to challenge the decisions made by the algorithms (Calvo et al., 2020; Kahn, 2020; AlgorithmWatch, 2020; Milan and Treré, 2020). Even voluntary signing up to data-driven software can lead to loss of privacy. For example, in the case of Google's Verily symptom tracking software, people who wanted to have their COVID risk assessed needed to already possess or make a new Google account and enter their personal details, thus raising concerns about what Google/Verily would do with their information. Another difficulty is the potential for the datasets and algorithmic processing used to calculate COVID-19 risk to unfairly confine people to isolation and allow them no opportunity to challenge the decision made by the software. Examples of such inaccuracies have been reported by Chinese citizens subjected to the COVID health rating app. As one man who had erroneously been rated as 'red' claimed: "'I felt I was at the mercy of big data", . . . "I couldn't go anywhere. There's no one I could turn to for help, except answer bots'" (Zhou, 2020).

In the UK, the 'shielded patient list' incorporated sensitive data about a person's ethnic identity, their likely experience of socioeconomic disadvantage by location, their age, weight and existing health conditions to algorithmically predict to what extent people are clinically vulnerable to severe complications from COVID infection. People who had been identified as 'at risk' were placed on the list and informed by government authorities that they needed to shield (protect themselves) from exposure to infection. The implications of being included on the list were that people were expected to severely limit their movements outside their homes, which could involve further risks: to their mental health, economic status and social relationships. For those who were wrongly identified as 'high risk', the consequences were social exclusion, but those who are wrongly not included on the list could expose themselves to a heightened risk of COVID by not shielding themselves and missing out on financial and healthcare support. The accuracy of the algorithm, therefore, had significant implications for people's health and wellbeing, including potentially exacerbate existing health problems and socioeconomic disadvantages (Patel, 2021).

Data-driven systems for identifying who should have priority for COVID vaccinations have similarly been subjected to criticism. One example is the Stanford Medical Center, USA, whose vaccination decision-making algorithm was publicly challenged in December 2020 for sending frontline healthcare workers to the back of the vaccination queue, while hospital administrators who had no contact with patients received priority. It was revealed that the problem with the algorithm was that it was weighted too heavily towards prioritising staff members who were older, using age and theoretical risk of infection, rather than incorporating data about staff members' job duties, direct contact with the novel coronavirus and actual risk of infection, injury, illness and death. Such errors in ADM software highlight the hierarchical nature of hospital systems in the USA and elsewhere, in which frontline workers (particularly low-ranked workers such as cleaners and orderlies) are considered more expendable than physicians and administrators (Cabrera, 2021).

Surveillance technology companies have stepped in to offer services that appear to solve the 'problem' of the pandemic in the interests of public health. New systems have been introduced with little warning or public scrutiny and debate. These new services have included rebranding cameras designed to detect weapons as thermal scanners supposedly to detect COVID infections and the use of drones to detect people not conforming to physical distancing or quarantine regulations. Major national border controls and surveillance, including along the US–Mexican border, the Myanmar–Thai border and in the Mediterranean region, have begun to use COVID as an excuse for introducing greater automated systems such as facial recognition to prevent entry of migrants and refugee-seekers (Venkataramakrishnan, 2021; Gershgorn, 2021).

These and other examples of function creep require sustained examination for their implications for human rights. In some countries, human rights activist organisations have protested against such technologies, with some success. In Paris, civil liberties groups La Quadrature du Net and the Human Rights League took the Paris police service to court to protest against the use of drones equipped with cameras to monitor residents' compliance with COVID restrictions. This lawsuit was successful, with the court banning this technology (Fouquet and Sebag, 2020). In another Paris example, a trial of the use of cameras in one of the busiest subway stations in that city to detect whether passengers were wearing masks was soon halted because activists pointed out that it contravened the EU general data protection regulation. In the UK, the Information Commissioner Elizabeth Denham made a statement in October 2020 criticising data brokering companies who were seeking government support to process Britons' personal data to identify people who may be at 'high risk of breaking self-isolation' during lockdown periods (Venkataramakrishnan, 2021).

The success of such initiatives, however, relies on the opportunity for powerful activist groups or privacy advocates to take action and also on the extent to which COVID dataveillance is visible and knowable. In other situations, people lack awareness of the technologies and their implications. The Centre for Data Ethics and Innovation conducted a longitudinal survey of UK citizens for six months (between June and December 2020) to elicit their attitudes towards the use of AI for COVID management. The survey findings were revealing in terms of the lack of knowledge the British respondents demonstrated in terms of both general issues concerning their personal data privacy and protection and the digitised COVID control and surveillance strategies that had been implemented or anticipated in the UK. The survey recorded high support on the part of the British public for the potential of digital technology to combat COVID, which remained consistent throughout the six months the survey was repeated. These findings suggest that the logics and promissory narratives concerning digital health technologies that had been promoted in the UK and elsewhere for COVID control were accepted unquestioningly by many people, as they were viewed as being for the public good.

In the case of disempowered minority groups such as refugees, illegal immigrants or people living in countries with authoritarian governments where human rights are rarely championed or upheld, there is little opportunity for citizens to challenge invasive or discriminatory COVID dataveillance. Indeed, many impoverished communities – particularly in the Global South – suffer from the problem of 'data poverty' and invisibility because they are marginalised or ignored completely by datafication and dataveillance systems. Undocumented migrants are unlikely to come forward to be included in dataveillance or digitised care systems such as vaccine delivery because they face the risk of deportation. Most digitised COVID surveillance, control and care systems – even the most simple, such as using QR codes to check-in to locations – lack inclusivity by their very design. This software tends to rest on the assumptions that people are privileged enough to own recent-model smartphones, have reliable access to Wi-Fi and possess the knowledge and language skills to successfully use these technologies (Milan and Treré, 2020; Milan, 2020).

Discussion and concluding comments

As is so common with many other representations of ADM, data-driven systems and associated technologies applied to health problems and crises (Lupton, 2014, 2017b), the logics, promissory narratives and sociotechnical imaginaries of digitised COVID surveillance and control have presented a techno-utopian portrayal, in which these technologies are positioned as offering more effective and efficient pathways to managing the COVID crisis. In a sociocultural context in which interactions with the fleshy and potentially virally contaminated bodies of other people have been continually problematised as life-threatening, the remoteness and hygienic technological imaginaries associated with digitised COVID technologies promise to offer greater safety and security. The affordances of digital technologies to generate real-time metrics for dataveillance and to facilitate arms-length and self-administered diagnosis and medical care have been emphasised in the discourses promoting these technologies.

As I have demonstrated in this chapter, in some ways, data-driven digitised strategies for COVID surveillance, care and control have fulfilled these promises.

However, many proposed or implemented digitised 'solutions' to the problem of COVID have had little effect or have fallen by the wayside, despite overblown claims for their problem-solving capacities. Others have been implicated in exacerbating existing socioeconomic disadvantage and social discrimination or flouting privacy rights. Most digital systems have been designed with socioeconomically privileged users from the Global North in mind, thereby ignoring the specific needs and unintended consequences for less privileged communities and populations. Global public health and medical responses to the COVID crisis have combined ages-old restrictive and authoritarian state governance of populations with more recent neoliberal imperatives to conform to expectations for people to be self-responsibilised, morally upstanding citizens who are willing to engage in risk-protective behaviours: for both their own good and the good of society at large.

Critical scholars have been quick to identify COVID surveillance management policies implemented by governments as a prime example of Foucauldian biopolitics, disciplinary power and government by populations by emphasising voluntary self-management of risk (e.g. Bennett, 2021). To some extent, however, as I have shown, many digital and non-digital strategies for COVID management rely on the direct operation of sovereign power by those in positions of authority enforcing strategies such as border closures, stay-at-home/sheltering in place restrictions, closures of business, government agencies and schools, unprecedented access to telecommunications databases and imposing harsh fines if these restrictions are flouted. In many jurisdictions, people who have been ordered to go into quarantine or self-isolation have been closely monitored by in-person checks undertaken by health authorities or police officers or with the use of digitised dataveillance. In some cases, they have been subjected to public 'naming and shaming' as well as fines, with their personal details such as their names, addresses and occupations openly revealed. This is particularly the case for countries ruled by authoritarian governments, with China among the most prominent, but these practices have also been implemented in liberal democracies such as Australia, the UK and the USA to an extent that is unprecedented in modern times.

My analysis of digitised COVID technologies has highlighted the lack of contestation over the promissory narratives and sociotechnical imaginaries that have characterised the development, promotion and deployment of these devices and software. This exercise of sovereign power has, with only some exceptions, received little protest or challenge on the part of publics, because it is viewed as an appropriate and necessary emergency response to the crisis. While activist organisations and critical scholars have attempted to sound a note of caution, pointing to the potential to flout human rights to privacy or reinforce stigmatisation and socioeconomic marginalisation that are part of many of these initiatives, techno–utopian visions inserted within a prevailing politicised atmosphere of crisis have trumped or excluded these considerations and caveats.

Throughout the pandemic, concepts of care that are directed at protecting the health of the population have been privileged over those that focus on enhancing and facilitating basic human rights or protecting already disadvantaged social groups. Indeed, some commentators have argued that the COVID crisis has surfaced a 'crisis of care', in which the failings of neoliberal political and privatised approaches to public health surveillance systems and healthcare delivery across the world have been suddenly revealed (The Care Collective, 2020). Digitised systems of health surveillance and symptom checking involving ADM have been introduced in part to deal with this crisis of care. As I have argued, however, they do so in ways that often display a fundamental lack of care for the needs and human rights of those who are incorporated within these more-than-human assemblages.

Underpinning the introduction of ADM and data-driven strategies for COVID control and management purposes is the assumption that automated public health surveillance is superior to traditional approaches and that some people or social groups cannot be trusted to self-regulate and, therefore, must be placed under dataveillance regimes. Current digitised methods of control, containment and care harken back to anxieties about contaminated and uncontained – and therefore risky – human bodies and their movements across place and space. We can see in these discourses and practices a potent combination of old and new ways of controlling infectious disease outbreaks which requires continuing examination and analysis if ages-old forms of social discrimination and neglect are not to be reproduced and reinforced even further. Rather than acknowledge the interdependence of people with each other, with other living things and with non-living things as part of caring relationships and connections (The Care Collective, 2020; Milan, 2020), such approaches position people as responsibilised individuals requiring close monitoring and disciplining. Not only are there ample opportunities for people and social groups who are already marginalised and disadvantaged to be unfairly targeted by discriminatory algorithms and ADM for COVID control and surveillance, but the potential for the 'data poor' to be excluded from healthcare, vaccination programs and financial support must also be identified and challenged.

References

ABC News Online (2020) Donald Trump Says He Ordered Slowdown in Coronavirus Testing in Speech to Rally in Tulsa, Oklahoma. *ABC News Online*. Available at: www. abc.net.au/news/2020-06-21/donald-trump-says-ordered-slowdown-coronavirus-testing/12377556

AlgorithmWatch (2020) *Automated Decision-Making Systems in the COVID-19 Pandemic: A European Perspective*. Available at: https://algorithmwatch.org/en/automating-society-2020-covid19/

Armstrong D (1995) The Rise of Surveillance Medicine. *Sociology of Health & Illness* 17(3): 393–404.

Bashford A (ed) (2016) *Quarantine: Local and Global Histories*. Houndmills: Palgrave Macmillan.

Bennett JA (2021) Everyday Life and the Management of Risky Bodies in the COVID-19 Era. *Cultural Studies* 35(2–3): 347–57.

Cabrera JA (2021) Rewriting the Vaccine Distribution Algorithm. *Points – Data & Society*. Available at: https://points.datasociety.net/vaccinating-health-care-workers-c4372dc731de

Calvo RA, Deterding S and Ryan RM (2020) Health Surveillance During Covid-19 Pandemic. *British Medical Journal* 369. Available at: www.bmj.com/content/bmj/369/bmj.m1373.full.pdf

Centre for Data Ethics and Innovation (2021) COVID-19 Repository and Public Attitudes: 2020 in Review. Available at: https://assets.publishing.service.gov.uk/government/uploads/system/uploads/attachment_data/file/967585/CDEI_COVID19_Repository_and_Public_Attitudes.pdf

Douglas M (1992) *Risk and Blame: Essays in Cultural Theory.* London: Routledge.

Elish MC and Watkins EA (2020) *Repairing Innovation: A Study of Integrating AI in Clinical Care.* New York: Data & Society Institute.

feevr.tech (2021) *Feevr.* Available at: https://feevr.tech/

Foucault M (1984) The Politics of Health in the Eighteenth Century. In: *The Foucault Reader.* New York: Pantheon Books, 273–89.

Foucault M (1991) Governmentality. In: Burchell G, Gordon C and Miller P (eds) *The Foucault Effect: Studies in Governmentality.* Hemel Hempstead: Harvester Wheatsheaf, 87–104.

Fouquet H and Sebag G (2020) French COVID-19 Drones Grounded After Privacy Complaint. *Bloomberg.com.* Available at: www.bloomberg.com/news/articles/2020-05-18/paris-police-drones-banned-from-spying-on-virus-violators

Gershgorn D (2021) Covid-19 Ushered in a New Era of Government Surveillance. *One Zero.* Available at: https://onezero.medium.com/covid-19-ushered-in-a-new-era-of-government-surveillance-414afb7e4220

Goh B (2020) China Rolls Out Fresh Data Collection Campaign to Combat Coronavirus. *ITNews.* Available at: www.itnews.com.au/news/china-rolls-out-fresh-data-collection-campaign-to-combat-coronavirus-538635

Graham H and McIvor G (2017) Electronic Monitoring in the Criminal Justice System. *Iriss.* Available at: www.iriss.org.uk/resources/insights/electronic-monitoring-criminal-justice-system

Hacking I (1990) *The Taming of Chance.* Cambridge: Cambridge University Press.

Jasanoff S (2015) Future Imperfect: Science, Technology, and the Imaginations of Modernity. In: Jasanoff S and Kim S-H (eds) *Dreamscapes of Modernity: Sociotechnical Imaginaries and the Fabrication of Power.* Chicago, IL: University of Chicago Press, 1–33.

Kahn JP (2020) *Digital Contact Tracing for Pandemic Response: Ethics and Governance Guidance.* Baltimore, MD: The Johns Hopkins University Press.

Lupton D (1994) *Moral Threats and Dangerous Desires: AIDS in the News Media.* Bristol: Taylor & Francis.

Lupton D (1995) *The Imperative of Health: Public Health and the Regulated Body.* London: Sage.

Lupton D (2013) *Risk.* London: Routledge.

Lupton D (2014) Beyond Techno-Utopia: Critical Approaches to Digital Health Technologies. *Societies* 4(4): 706–11.

Lupton D (2016) *The Quantified Self: A Sociology of Self-Tracking.* Cambridge: Polity Press.

Lupton D (2017a) Digital Bodies. In: Andrews D, Silk M and Thorpe H (eds) *Routledge Handbook of Physical Cultural Studies.* London: Routledge, 200–8.

Lupton D (2017b) *Digital Health: Critical and Crossdisciplinary Perspectives.* London: Routledge.

Lupton D (2019) Toward a More-than-Human Analysis of Digital Health: Inspirations from Feminist New Materialism. *Qualitative Health Research* 29(14): 1998–2006.

Lupton D (2021) Contextualising COVID-19: Sociocultural Perspectives on Contagion. In: Lupton D and Willis K (eds) *The COVID-19 Crisis: Social Perspectives.* London: Routledge, 14–24.

Mack A (1991) *In Time of Plague: The History and Social Consequences of Lethal Epidemic Disease.* New York: New York University Press.

Martin R (2006) The Exercise of Public Health Powers in Cases of Infectious Disease: Human Rights Implications. *Medical Law Review* 14(1): 132–43.

Milan S (2020) Techno-Solutionism and the Standard Human in the Making of the COVID-19 Pandemic. *Big Data & Society* 7. https://doi.org/10.1177/2053951720966781

Milan S and Treré E (2020) The Rise of the Data Poor: The COVID-19 Pandemic Seen from the Margins. *Social Media + Society* 6. Available at: https://doi.org/10.1177/2056305120948233

Mol A (2008) *The Logic of Care: Health and the Problem of Patient Choice.* London: Routledge.

Olle E and AAP (2020) Coronavirus Drones Developed by University of South Australia Researchers. *7news.com.* Available at: https://7news.com.au/lifestyle/health-wellbeing/coronavirus-drones-developed-by-university-of-south-australia-researchers-c-763510

Patel R (2021) Why the COVID-19 Shielded Patient List Might Both Compound and Address Inequalities. *Ada Lovelace Institute.* Available at: www.adalovelaceinstitute.org/blog/covid-19-shielded-patient-list-inequalities/

Raley R (2013) Dataveillance and Countervailance. In: Gitelman L (ed) *"Raw Data" Is an Oxymoron.* Cambridge: MIT Press, 121–45.

Rhodes T, Lancaster K and Rosengarten M (2020) A Model Society: Maths, Models and Expertise in Viral Outbreaks. *Critical Public Health* 30(3): 253–6.

Spence A (2021) Grounded Pandemic Drone Earns Second Chance in the US. *The Lead.* Available at: https://theleadsouthaustralia.com.au/hi-tech/grounded-pandemic-drone-earns-second-chance-in-the-us/

Sutton M (2020) Phone Tracking Used to Follow Movements of Chinese Couple with Coronavirus in Adelaide. *ABC News Online.* Available at: www.abc.net.au/news/2020-02-06/phone-tracking-follows-movements-of-couple-with-coronavirus/11935912

The Care Collective (2020) *The Care Manifesto.* London: Verso.

van Dijck J (2014) Datafication, Dataism and Dataveillance: Big Data Between Scientific Paradigm and Ideology. *Surveillance & Society* 12(2): 197–208.

Venkataramakrishnan S (2021) Algorithms and the Coronavirus Pandemic. *Financial Times.* Available at: www.ft.com/content/16f4ded0-e86b-4f77-8b05-67d555838941

Wernimont J (2019) *Numbered Lives: Life and Death in Quantum Media.* Cambridge: MIT Press.

White L and Van Basshuysen P (2021) Without a Trace: Why Did Corona Apps Fail? *Journal of Medical Ethics* 47(12): e83. Available at: https://jme.bmj.com/content/47/12/e83.abstract.

Winter L (2020) Data Fog: Why Some Countries' Coronavirus Numbers Do Not Add Up. *Aljazeera.* Available at: www.aljazeera.com/features/2020/6/17/data-fog-why-some-countries-coronavirus-numbers-do-not-add-up

World Health Organization (2021) WHO Coronavirus (COVID-19) Dashboard. Available at: https://covid19.who.int/

Zhou V (2020) How Big Data Is Dividing the Public in China's Coronavirus Fight – Green, Yellow, Red. *South China Morning Post.* Available at: www.scmp.com/news/china/society/article/3051907/green-yellow-red-how-big-data-dividing-public-chinas-coronavirus

4

LESS WORK FOR TEACHER? THE IRONIES OF AUTOMATED DECISION-MAKING IN SCHOOLS

Neil Selwyn

Introduction

As in many areas of society, educational institutions are beginning to adopt all manner of automated decision-making (ADM) technology. While educational applications and systems have not tended to feature in critical accounts of artificial intelligence (AI) and automation, they offer rich examples of emerging impacts (and tensions) associated with ADM in everyday social settings. This chapter examines one such example of an ADM system marketed recently to Australian schools. This product promises to automate the registration of students' in-class attendance – what is referred to in Australian schooling as the 'roll call'. Instead of teachers reading through alphabetical lists of names and deciding which students are present (and which are not), schools can now invest in 'AutoRoll' – a facial recognition 'solution' for classroom attendance management. As we shall see, this seemingly innocuous system foregrounds a range of ways in which initial aspirations of ADM designers and developers can bump up against context-specific practices and understandings. All told, there is a lot more to ticking students' names from a list than might be first assumed.

A brief overview of classroom ADM

There is now a steady growth of ADM technologies designed for educational use. Perhaps most prominent are technologies intended to support student learning, such as personalised learning systems designed to direct students' engagement with online learning resources. These systems use sophisticated data-driven analytics to support decision-making regarding what students should be learning next. Perhaps more prevalent are administrative and organisationally focused forms of AI-driven technology – mostly designed to support routine automated decision-making for institutions, teachers and other staff. Many of these technologies support what

DOI: 10.4324/9781003170884-6

Gulson and Witzenberger (2021) term 'automated education governance' – ranging from in-school bureaucratic decision-making to the management of national education systems. Elsewhere, AI-driven technologies have been adopted by schools to support the initial stages of teacher recruitment, purchasing and resource procurement, predicting patterns of likely student enrolment and retention, and informing various other 'business decisions' faced by school administrators.

Alongside these institutional forms of ADM, a number of other AI-driven technologies have been developed to support classroom decision-making tasks that previously would have fallen to teachers. These tasks range from judging the quality of student work through to identifying students who cheat, or perhaps are de-motivated and disengaged from their studies. For example, there is growing interest in the use of AI systems which monitor students' attention levels and emotional states. Popular essay plagiarism tools such as 'TurnItIn' are now bolstered by the use of AI-based 'language stylometrics' to assist teachers in deciding on instances of academic malpractice and cheating. School systems and individual schools are also beginning to make use of automated test-scoring and essay assessment (what is sometimes described as 'robo-grading') to support grading decisions (Shermis and Lottridge, 2019).

Yet, perhaps the most pervasive form of ADM being taken up by schools are technologies that support 'gatekeeping decisions' – managing the flow of people in, through and out of school spaces. Most prosaically, this has seen the rise of automated 'visitor management systems' to support the signing-in process of students, staff and other on-campus personnel and visitors. The rise of facial recognition and object detection technology has seen the adoption of facial recognition systems by thousands of US schools in efforts to identify unauthorised and potentially harmful intruders. During the return to face-to-face schooling after COVID shutdowns, reports surfaced of US schools using pandemic relief funding to purchase new facial recognition systems, with intruder detection, attendance monitoring and added thermal imaging capabilities (Barber, 2020). Similar biometric systems are being used to authenticate the identity of online students – in other words, confirm that the people engaging in off-campus online learning activities are actually who they claim to be. In many ways, having a clear idea of who is 'in attendance' at any particular time is a prerequisite to anything else taking place.

The promises and logics of educational ADM

The claimed benefits associated with these different technologies will be familiar to general observers of ADM. For example, commonplace promises of ADMs leading to greater precision are evident in hopes of bringing a formal mathematical logic of 'calculability' to bear on classroom processes that are traditionally guided by informed teacher guesswork or speculative planning. Similarly, promises of ADMs leading to enhanced 'insight' and informed action are evident in hopes of the technology-driven 'customisation' and 'personalisation' of education provision (see Bulger, 2016) – therefore, countering criticisms that 'cookie-cutter' mass schooling has proven unable to cater for diverse individual needs.

This idea of ADM acting as a corrective to specifically educational shortcomings is perhaps most evident in promises of ADMs leading to greater efficiencies by reducing (or removing) the number of 'humans in the loop'. Here, such discourses are often framed in terms of ADM technology acting as a corrective to teacher frailties, such as fatigue, bias and scarcity of attention to individual students. Indeed, while ADM in all walks of life is justified along lines of 'avoid[ing] the biases, prejudices and irritations of human [actors]' (Lisle and Bourne, 2019: 682), this is seen to be a particular consideration when it comes to teachers dealing with large numbers of students. School teachers around the world are acknowledged as over-worked, worn-out and burnt-out (Fitzgerald et al., 2019; Saloviita and Pakarinen, 2021). In response, AI-driven systems are imagined as driven by dispassionate, objective and impartial decisions – what Edwards and Cheok (2018: 5) describe as 'the ability of machines . . . to interact with human learners without human emotions getting in the way'.

This is not to say that these technologies are envisioned as doing away with teachers altogether. Instead, such claims are usually tempered by a sense of ADM relieving teachers – 'freeing-up' teachers from routine and procedural tasks, allowing them to concentrate on higher-level pedagogic work. Mirroring broader claims of ADM 'freeing up' knowledge workers to engage in more complex discretionary work (McIntyre, 2019), AI technologies are described as taking responsibility for onerous classroom 'routines', 'duties' and 'drudge' work associated with teaching. As such, it is commonly argued that teachers might soon benefit from having AI-driven 'assistants' that provide 'intelligent support' and reduce workload and stress (Ideland, 2021). Such scenarios, therefore, anticipate:

> a future in which the role of the teacher continues to evolve and is eventually transformed; one where their time is used more effectively and efficiently, and where their expertise is better deployed, leveraged, and augmented.
>
> *(Luckin et al., 2017: 11)*

Implicit in the drive to integrate ADM into schools is a sense of schools being perceived with concern and bewilderment by those responsible for designing, developing and marketing the technology. In this sense, development of classrooms ADMs chimes with ideas of school institutions and systems that are woefully 'unfit for purpose' for digitally driven societies. The argument is often made, for example, that conventional schools are 'broken', 'nineteenth century' and frustratingly 'cookie-cutter' in nature and form – that is outmoded and obsolete products of a bygone era. Commentators speak with exasperation of 'industrial era classrooms', schools resembling factories, 'ivory tower' educators and so on. In tandem with these frustrations are ambitions to engineer the 'corporate reform' of educational institutions – in short, the imposition of business models, logics and processes on schools and those who work in them (see Selwyn, 2021). ADM, therefore, fits well with wider appetites for technology-driven reforms that imbue business management ideals of efficiency, effectiveness and standardisation. Thus, any promises of 'freeing-up' and 'assisting' teachers are cut through with a sense of ADM adding a dose of much-needed corporate backbone to failing public institutions and rapidly fading workforces.

The case of 'autoroll' automated classroom attendance monitoring

We now go on to consider a modest example of the rise of school-based ADM – the development by a small Australian start-up of an 'automated attendance taking' system that uses facial recognition technology to automatically mark the class register at the beginning of each lesson. Here, we examine this technology through a close reading of patents, developer blogs, marketing materials and a series of interviews with key actors across the education sector. These empirical data provide insights into this particular ADM technology's technical functions, broader organisational logics, and ways in which it is understood to (re)shape classroom actions and relations.

Technical arrangements

In the dry terms of its product patent, AutoRoll (pseudonym used) is 'localised self-learning for recognising individuals at locations'. In more shrill marketing terms, AutoRoll promotion video depicts it is a 'revolutionary technology that automates roll call . . . it does automatically what school staff do manually'. All told, AutoRoll is best described as a facial recognition system designed to automate a tightly bounded moment of decision-making that takes place at the beginning of each school lesson – that is recording which students are in attendance. As one of the start-up founders describes:

> It is a very specific outcome that we are looking to solve which is just being able to say, "Who is in this room at this time? Okay, these students are expected to be here, they are". So let's just automatically mark that roll for the teacher. They can then log in, say "Yes, I can see all these data points are correct, I will submit that roll".

Initial iterations of the product involved wall-mounted iPads and an 'off the shelf' cloud-based facial recognition service. However, the company has moved quickly to a complex arrangement of self-contained custom-built hardware and 'software as service'. As such, AutoRoll now involves the classroom installation of small wall-mounted devices that streams video from 4K cameras. In addition, the company's patent application describes a 'visitor management' version integrated into a kiosk located in reception areas which 'may also be interfaced with other external physical devices to provide access control, such as controlling the magnetic lock of a door'.

Both these devices are configured as 'things' in the Internet of Things (IoT). This means that all communication, authentication and security take place through the IoT framework in Amazon Web Services (AWS). The co-founder and technical lead of AutoRoll adds that the company has developed its own 'purpose-built' real-time models which are stored onboard each device for 'extracting the points of interest, doing a quick identification, and then, if it can't be figured out or we haven't seen that person before, off to Rekognition [Amazon's computer vision cloud service] we go to get that final answer'.

Framing the 'educational' problem

These technical specifications reflect distinct framings of the real-life 'problems' being addressed that AutoRoll founders have pursued since the inception of their product. In the solutionist rhetoric of start-up culture, one favoured 'elevator pitch' for the product is: 'how can we solve people management and attendance at schools?'. Interestingly, when addressing IT industry audiences, AutoRoll founders are keen to play down the 'bland' nature of their task-of-choice. In this sense, automating the roll is justified as relieving schools of a boring but mandated aspect of schooling, a point of 'compliance' and something that 'schools are legally required'.

Nevertheless, also evident is what AutoRoll's developers identify as a problem of scale:

> When you think about it, schools are a very complicated problem. We've got the teachers needing to mark the roll at least every hour . . . But then you've got the issue of visitor management and compliance around there as well. Schools have got people visiting their campuses all day every day and we need to verify who those people are . . . if you think about the size of schools, this is really a problem of scale . . . a lot of these things are easy to do manually when you've got a small number of students or a small number of classrooms, but when you introduce scale to that problem, it just gets harder and harder and harder.

Elsewhere, AutoRoll's developers describe schools as places with 'about 10–20% of their population who are either absent or sign in late or leave early'. Drawing on terminology from transportation management and webpage design, this is seen to cause a problematic escalation of 'dwell time'. All told, from a software engineering point of view, 'it's quite a large amount of data that you've got to process'.

Overcoming 'technical' challenges

Indeed, AutoRoll's main technical concern had initially been conceived in terms of this 'problem of scale'. The small development team saw large classrooms as requiring at least three 4K cameras in order to get full visual coverage at sufficient pixel density 'needed to do a good recognition event'. This meant that AutoRoll's imagined school ('with say 80 – even 100 – classrooms') produces large streams of data. This volume of data traffic was seen as the primary technical challenge:

> You can imagine that we have hundreds and hundreds of devices at a customer's site running during the day. They're firing data at our cloud relentlessly.

> We're horizontally scaling out as big as we can, and then the clock hits 4
> o'clock . . . everyone goes home and everything goes dead. It's all quiet.

However, the company's initial piloting foregrounded an additional problem of
school expectations around children and data privacy. The product's pilot phase
quickly revealed that any school ADM is subject to a number of regulatory frameworks – various state 'Privacy & Data Protection' acts, the 'Federal Privacy' act, and
even niche regulation such as a 'Surveillance Devices' act. As an educational legal
advisor from the Independent Teaching Union puts it, the application of this legislation in schools was 'pretty broad' and 'not well tested'. As such, AutoRoll's initial
forays into schools prompted considerable push-back from local policy makers,
parents and media – even prompting one state government Education Minister to
declare a temporary moratorium on the implementation of any facial recognition
technology in government schools. As AutoRoll's UX research put it:

> Parents and teachers feel uncomfortable and concern with the idea of having
> an AI-based attendance system in the educational space as well as privacy
> governance.

Issues of privacy, therefore, formed a key subsequent technical problem for the
AutoRoll team to address – 'this is where we spent a lot of engineering time and
a lot of effort'. This involved AutoRoll products being redesigned to be completely self-contained – with AI algorithms running on specifically manufactured
hardware devices which did not produce or store images of people or whole-class
activities. Ironically (given marketing claims of teachers being overwhelmed by
administering large classes) AutoRoll's product patent describes this as a modest
data-load – 'processing by the local device remains fast and computational (sic)
efficient, as it only needs to remember a small group of people'.

Technically this is a difficult and profit-limiting approach to take – not least
because manufacturing their own hardware restricted export opportunities as compared to solely selling software. Nevertheless, the founders are keen to be able to
claim that their products 'can't be misused. Even innocently, they can't be misused'. In this sense, complex societal debates around facial recognition, ethics and
discrimination that peaked while AutoRoll was coming to market were able to
be tamed in bounded terms of schools' 'legal requirement' to address matters of
'privacy'. This was a 'requirement' that AutoRoll could claim to have addressed
through its custom-made technology, as described by the co-founder:

> It is a lot more expensive in terms of up cost to actually get hardware prod-
> ucts rolled out . . . but we realised it was an absolute requirement . . . we went
> down the path of doing that R&D because it is the requirement for schools.
> Other companies might be looking at using AI and facial recognition for
> broader use-cases and scopes, but for us it was very much focused on how

is this appropriate, and how is this ethical for the use, and comply with legal requirements for use in schools?

Perceived benefits

Despite the 'specific' nature of the decision-making task, this considerable technical work underpinned various exuberant claims of AutoRoll 'helping schools'. First, are familiar claims of efficiency – 'we've developed the solutions to streamline and improve the efficiency of [attendance] processes for schools'. These efficiencies are described in terms of precision ('accurate attendance records') and time-saving ('instant roll call', 'real-time knowledge', and overcoming the problem that 'those few minutes add up and can equate to two and a half hours of lesson time lost per week for every student', as summarised in the AutoRoll Frequently Asked Questions.

Second, are associated benefits of 'eliminat[ing] human error'. The system is framed as 'completely foolproof', even when faced with students dishonestly attempting to register absent classmates. Similarly, teachers are framed as fallible – 'even with the best of intentions, mistakes can happen'. As AutoRoll's patent put it, 'manual attendance tracking is labour-intensive, time-consuming, and prone to circumvention and inaccuracy'. Taking the class roll is, therefore, framed by the start-up founders as an unreasonable task to expect teachers to undertake manually:

> When you're asking a teacher . . . to check 20–25–30 names off a list once an hour of every day, while keeping a classroom engaged, while not getting distracted because someone's walking in ten minutes late, while trying to get a lesson running and not being able to just stop and then log into a computer and check off the digital register, which might take two or three minutes, and by that time the classroom gets rowdy again and they start throwing paper planes at each other and then it takes another five minutes to get their lesson on track. That happens all the time, and therefore there is human error. There's always human error around that attendance taking process.

In this sense, the use of facial recognition is justified as a failsafe means of managing these potential risks and ensuring a provision of safety. Indeed, over time Auto-Roll's marketing has increasingly promoted issues of care and safety – 'AutoRoll's all about helping organisations deliver great care to their communities . . . provid[ing] environments that all people can feel safe and trust'. In the words of the company founders and wider biometrics industry, anticipating issues of care are framed in extreme terms:

> No parent wants to drop their child off at school in the morning and have a thought that the school doesn't know where their child is during the day or, God forbid, even lose them.

> If there's high risk of bad people being in the area of the school and you
> want to protect your children, then [with AutoRoll] you might start feeling
> more comfortable.

All told, AutoRoll was justified as fulfilling a public service – reflecting best industry principles of 'ethical AI' and 'AI for good' by selflessly focusing on the explicit task of classroom attendance, rather than more 'powerful' AI-driven impacts. As described by the co-founder:

> It comes down to an ethical view of technology development. . . . Realistically, AI is such a powerful tool that can be deployed to solve so many different problems. . . . We get approached quite frequently from different potential customers, asking us to produce different aspects of our AI that could have really positive impacts to do with what our attendance taking solutions can do. Things like being able to support with potential mental health effects, depression, anxiety. There are off-the-shelf algorithms that can be used to do that. . . . But then your consideration needs to be around . . . is it going to be used in the right way? So, we take all of that into consideration, and right now we say 'No, we're not going to do that'. Because we haven't done enough analysis, and we haven't done enough of a product scope to actually understand what the specific outcomes are to make sure that then the technology is used for good? So, it's AI for good and not with the potential to be detrimental or biased.

Feared diminishments

Finally, then, it is worth contrasting these commercial framings of educational 'problems' and automated 'solutions' with the views of education professionals. Despite its commercial bluster, the AutoRoll system has been taken up only in a handful of schools – with the company remaining tight-lipped about its customer-base. This is understandable, given the scepticism that we noted in our interviews with various key stakeholders from across the education community. The majority of these respondents were notably underwhelmed by the idea that using facial recognition technology to automate the roll-call addressed a notable problem. This included observations by the legal advisor that 'it can be pretty well assumed that most schools will know exactly who's on the premises', and the chair of the 'biometrics in schools' lobby group, who had 'not heard of any school that couldn't run without a biometric system'. All told, AutoRoll was generally seen as a disproportionate use of a controversial technology on a vulnerable population. As one government data privacy official concluded: 'there is no necessity in the end for the use of [facial recognition] – the balance isn't right'.

Key here, then, was a sense that AutoRoll was predicted on an 'outsider' view of schools and classrooms – 'it depends on whether you're looking at schools from the outside or the inside'. Doubts were raised in a group interview of the executive and teacher representatives of the national teaching union that the technology would operate smoothly – incurring additional work for teachers: 'we know who the people are who are going to have to reboot it, reset it, all of that sort of stuff.

That's just another job for somebody in the school to do'. Perhaps most significant was this concluding discussion from a group of teacher representatives and trade union officials – expanding on the significance of the roll-call in the context of teachers' professional judgement and classroom management:

A: The assumption that taking the roll is lost time, and the assumption that the roll is not teacher work is a false assumption.

B: Yes, it's a pretty perfunctory process, but as a teacher you can use it in a whole range of ways that are actually about ensuring that the students are in the best place to learn, is focused. It is a commencement activity that has all the norms that sit around it. It's an institutionalised activity, and in that context teachers either use it in a very procedural way – get through it, get it done, move on . . . [but] other teachers use it in a very creative way as a tool to get students ready to learn.

A: Sometimes you do get it wrong, and it can be a bit annoying, you can stuff it up . . ., but at the same time it actually has a ceremonial process that can be very very useful to a teacher.

C: . . . and the interactivity that it creates between you and the students.

A: . . . but also between the students in a classroom. Because they are hearing the other students' names so they're actually being reminded that it's not just a physical presence – it's the symbolism of a name. It's going to put them in a different sort of mindset.

B: And teachers will use it not just as a way to check if a child's in a classroom but also to actually greet the child – to welcome them. It does cue the other kids. And it also has that dynamic that when a student is absent the other kids will tell you, which might then lead to a conversation as to why that is the case.

D: What about the new teacher? For a new teacher, the roll is a crucial process to get to know who is there. With the repetition, by the end of term you do get to know their names.

B: Also, it's your first assessment of the day. Depending on how a student actually answers you that's the first indication of what you have to do in that classroom.

E: It is also a chance to show a bit of discretion, if a kid was having troubles and was persistently late you might opt to delay taking the roll until 10 minutes into the lesson so they don't show up as being late yet again. If the roll is taken automatically in the first minute, then you don't have that leeway.

C: It's around the belonging . . . that whatever stage a student is at, they are part of this class in a way that is equal to all the other students.

A: So to narrow [AutoRoll] to it being efficient to decision-making absolutely side-lines a whole bunch of the relational processes . . . and that can't be discounted.

Discussion

In contrast to a surgeon deciding whether to operate, or a pilot deciding how to land a plane, a teacher taking the classroom roll might appear a decidedly incon-sequential instance of professional decision-making. Yet, the example of AutoRoll highlights the significant subtle tensions that can arise when any form of ADM is introduced into a professional work context. Indeed, AutoRoll raises the contention that there is no such thing as wholly routine professional decision-making. While taking the roll is something that outsiders (such as AutoRoll's developers) understandably might see as a waste of professional expertise and time, this is a 'duty' that some education professionals nevertheless consider to be a significant element of their autonomy and authority. This echoes a tension that runs throughout studies of ADM in other contexts – that is attempts to automate 'practices that operators do not consider automatic' (O'Grady, 2021: n.p.).

While it might appear to be a prosaic application of AI technology, AutoRoll raises a number of broader issues relating to the nature of ADM-infused classrooms, alongside automations of professional work in general. First, it is worth considering how the imperative for AutoRoll arose from – and is sustained by – broader sociotechnical conditions that have come to define contemporary education. On its own, the promise of cutting-back on time taking to register the class is likely not enough to ensure the widespread take-up of this technology. Instead, AutoRoll only continues to be perceived as a commercially viable possibility (at least by its industry and federal government supporters) because it fits neatly with broader logics and conditions of contemporary school reform.

For example, this is technology that aligns well with the extensive existing digital infrastructure of schools. Contemporary schools are now laden with digital devices, sensors and CCTV cameras, along with the continuous dataveillance through school platforms, learning management systems and other classroom apps. Perhaps more significantly, AutoRoll also 'fits' well with a number of broader prevailing logics of school improvement that help it ride out any initial push-back from concerned parents and politicians. For instance, the prospect of 'disrupting' a clerical practice stretching back to the late 19th century articulates well with discourses of the outdated and 'broken' industrial era school, as well as discourses of the over-worked, over-stretched teacher.

All told, the slightly 'over-tooled' imposition of real-time facial scanning technology is a totem of corporate reform – bringing business-related logics and efficiencies to bear on school organisation through the imposition of tech-driven business solutions. In a broad sense, then, AutoRoll begins to give shape to a long-standing sociotechnical imaginary amongst school reformers – the 'smart school':

> embedded with complex sensor networks that regulate learning environments through context-aware building management systems. These systems are capable of collecting and processing continuous streams of biometric and environmental data from school buildings and their inhabitants, including data collected from fingerprint scanners, facial recognition software, surveillance cameras, movement sensors, light sensors, and wearable biosensing technologies.
>
> *(De Freitas and Rousell, 2020: 11)*

It is, therefore, understandable that education 'outsiders' continue to be prepared to lend credence to AutoRoll's commercially motivated claims of the 'problem' of the manual class roll and the associated benefits of its automation. However, wide of the mark they might be, claims of being able to save up to 2.5 hours a week of 'lost lesson time' clearly appeal to ambitions to instil business-like efficiencies into classrooms. Similarly, AutoRoll's promotion of anticipatory anxieties over 'losing children' and harms arising from an absence of 'care' also correspond with dissatisfactions over distracted, over-worked and inattentive teachers. As is the case with other forms of security-related ADM, these echoes 'private firms' reliance upon the conjuring of risk-laden futures' (O'Grady, 2021: n.p.) and their desire to promote technologies as an opposing force for 'public good'.

Yet, notwithstanding such hyperbole, this chapter also highlights significant tensions in terms of how the technology diminishes the value of monitoring student attendance. In short, the underpinning logics of AutoRoll could be seen as rooted in what Wajcman (2019) terms an 'engineering model' of classroom processes and teaching tasks. As is the case with any ADM technology, the codification of the class roll process results in a finite, bounded model that inevitably constitutes 'a simplistic view of the tasks it supports and the structures it represents' (Fischer and Wunderlich, 2021: 5770). On the one hand, AutoRoll's design configures the act of taking a class roll as educationally unproductive and, therefore, inefficient human work that it is desirable to eliminate. On the other hand, this clearly contrasts with many teachers' actual enactment of this task in ways that are pedagogically and/or socially generative.

Indeed, the 'engineering mindset' framing of AutoRoll does not strictly consider the roll to be a 'decision' at all. AutoRoll presumes an objective act of recording – a student has simply entered the room or they have not – rather than a matter for discretion. Yet, an opposing sense arose from some of our teacher interviewees that what AutoRoll might codify as a simple and routine act should actually be acknowledged as complex and non-routine. Here, educators spoke of taking the classroom roll as a process of judgement, sense-making and assessing the situation. Crucially, then, calling the class roll can be a moment of considerable discretionary power based on professional judgement and expertise.

From a teacher's point of view, then, talk of automating the class roll needs to be set against a relational understanding of classrooms and teaching as relational work. In this light, the act of taking the roll is a key instance of tacit knowledge and social action, with each teachers' discretion when calling the roll enabling a significant moment of 'street level' classroom governance (McIntyre, 2019). Teacher discretion when calling the roll is necessary to make classrooms continue to run smoothly – smoothing over contradictions, allowing flustered children to calm down and late-arriving students a chance to recompose themselves. In contrast, the AutoRoll ADM curtails this professional judgement and relational work. Rather than 'freeing-up' teachers, this form of classroom ADM works to disintermediate their interactions with students, as well as diminishing the student as subject (e.g. describing a present child as a 'correct datapoint'). As such, AutoRoll's codification of the roll process sidelines the capacity of professional knowledge workers to 'tak[e] good decisions and solving problems, and that the abilities to do so derive from knowledge professional's individual creativity and deep experience' (Fischer and Wunderlich, 2021: 5770).

Conclusions

Much of the AutoRoll case reflects a familiar story of the realities of ADM integration into professional workplace settings. First is the idea that delegating responsibility for what seems to be a simple act of decision-making to ADM technology foregrounds the beliefs, views and logics of 'outsider' entrepreneurs, software developers and marketers responsible for the emergence of this technology. These delegations are not wholly unwelcome – rather they fit with a range of broader conditions of school reform, improvement and efficiency. Nevertheless, these technologies inevitably lead to a simplified codification of workplace processes which itself leads to 'the insertion of new rationalities and ontologies' into workplace settings (O'Grady, 2021). With teachers and students tending to be sidelined in this process, AutoRoll constitutes a subtle 'de-professionalising' presence in the working lives of educators. While ADM usually promises to overcome a number of human-related frictions, these frictions are a key element of the relational work that makes classrooms continue to function relatively smoothly on a day-to-day basis.

As such, the key conclusion to draw here is not really a matter of whether the continued development of ADM-driven roll-call technology is necessarily 'a good thing' or 'a bad thing' in schools. Rather, more thought needs to be paid to addressing (and perhaps reconciling) the differences in the ontological and epistemological grounds upon which the implementation of this technology is based (Lisle and Bourne, 2019). This chimes with a key theme running throughout this edited collection – that is how future integration of ADM technology might be reimagined along 'people-focused' and 'humanistic' lines. For example, what would a 'Auto-Roll' co-designed by teachers and students look like? What might ADM look like if it was designed to support teachers' relational work in classrooms? As Sarah Pink contends earlier in this volume, 'if we are to bring people into the process of ADM technology design, then we need to ensure that the conceptual categories that frame theory and practice in innovation account for people'. Therefore, a first step in reimaging rehumanised forms of AutoRoll might be re-examining the ways in which notions of 'care' are being used – that is the 'care' of automatically knowing a child has entered a room, as opposed to the 'care' of taking time to ask how the child is feeling. In this sense, for example, what might classroom ADM look like if it was designed around relational (rather than corporate) notions of 'care-full' work and 'caring' relationships, rather than notions of care as statutory duty?

Of course, it might be argued that refocusing ADM around the interests of teachers and students is not straightforward. While a number of possible inclusive and participatory directions might be co-opted from fields such as design, anthropology and human–computer-interaction, all of these run a risk of being thwarted by the current hegemonic conditions of educational institutions and educational work. Even after the educational upheavals of the COVID pandemic, the dominant model of compulsory schooling looks set to remain remarkably unchanged, with little appetite to switch over to online tuition, 'hybrid teaching' or similar technology-related shifts seen in the higher education sector. In particular, as Teräs et al. (2020) note, the 'post-pandemic' forms of educational technology that look set to dominate schools in the 2020s continue to be those that reinforce capitalist instrumental view of education rather than alternate values of promoting holistic human growth. In this sense, any efforts to involve teachers and students in the design, delegation and implementation of educational ADMs need to also address the challenges of rehumanising the broader institutional conditions and logics within which educational work takes place. In this sense, any re-imaginings of classroom ADM need to go hand-in-hand with broader educational reforms which strive to wholly re-establish schools as sites of cooperation and codetermination. This type of institutional renewal is a design challenge well beyond the scope of this chapter but should not be ignored in any attempt to imagine 'better' forms of educational ADM.

Acknowledgements

The chapter derives from fieldwork undertaken as part of two research projects – an Australian Research Council funded 'Discovery' project (DP200100189) and a preceding Monash University 'Inter-Faculty Grant'. The AutoRoll investigation is a collaboration with Liz Campbell and Mark Andrejevic.

References

Barber G (2020) Schools Adopt Face Recognition in the Name of Fighting COVID. *Wired*, 3 November. Available at: www.wired.com/story/schools-adopt-face-recognition-name-fighting-covid/

Bulger M (2016) Personalized Learning. Report, Data and Society, New York, July.

De Freitas E and Rousell D (2020) Relational Architectures and Wearable Space. *Research in Education* 107(1): 10–32.

Edwards B and Cheok A (2018) Why Not Robot Teachers? *Applied Artificial Intelligence* 32(4): 345–60.

Fischer L and Wunderlich N (2021) Datafication of Knowledge Work. In: Bui T (ed) *Proceedings of the 54th Hawaii International Conference on System Sciences*. Hawaii, 4–8 January. Honolulu: HICSS, 5769–78.

Fitzgerald S, McGRath-Champ S, Stacey M, Wilson R and Gavin M (2019) Intensification of Teachers' Work under Devolution. *Journal of Industrial Relations* 61(5): 613–36.

Gulson K and Witzenberger K (2021) Repackaging Authority. *Journal of Education Policy*. Epub ahead of print 11 October 2021. https://doi.org/10.1080/02680939.2020.1785552.

Ideland M (2021) Google and the End of the Teacher? *Learning, Media and Technology* 46(1): 33–46.

Lisle D and Bourne M (2019) The Many Lives of Border Automation. *Social Studies of Science* 49(5): 682–706.

Luckin R, HolMes W, Griffiths M and Forcier L (2017) *Intelligence Unleased*. London: Pearson.

McIntyre C (2019) Exploring Public Sector Managers' Motivations in Deploying Decision Support Tools to the Street Level. *Policy Quarterly* 15(1): 61–7.

O'Grady N (2021) Automating Security Infrastructures. *Security Dialogue* 52(3): 231–48.

Saloviita T and Pakarinen E (2021) Teacher Burnout Explained. *Teaching and Teacher Education* 97: 103221.

Selwyn N (2021) *Education and Technology*. London: Bloomsbury.

Shermis M and Lottridge S (2019) Communicating to the Public About Machine Scoring. Paper Presented to *National Council of Measurement in Education Annual Meeting*, Toronto, 7 April. Available at: www.air.org/sites/default/files/CSSC_Communicating%20with%20the%20Public_White%20Paper.pdf (accessed 11 October 2021).

Teräs M, Suoranta J, Teräs H and Curcher M (2020) Post-Covid-19 Education and Education Technology 'solutionism'. *Postdigital Science and Education* 2: 863–78.

Wajcman J (2019) The Digital Architecture of Time Management. *Science, Technology, & Human Values* 44(2): 315–37.

PART II

Embedding automated systems in the everyday

5

ALEXA'S GOT A HUNCH

The human decisions behind programming emotion-sensing and caregiving into digital assistants

Jenny Kennedy and Yolande Strengers

Introduction

Imagine that when you held your Apple iPhone to your face, Siri didn't just unlock your device but responded: 'You look anxious. Would you like to hear some soothing music?' and pulled a suitable playlist from your stored music. This kind of emotion detection with suggested 'solutions' is now possible. In 2016, Apple purchased Emotient, one of the leading companies focused on facial recognition utilising artificial intelligence (AI), and based on models that presume human emotions are uniform in the way they are expressed through facial expressions. Such models are controversial because they are premised on a disputed understanding that the way humans present emotions, or affect, is both innate and universal across cultures (Heaven, 2020). Facial and affect recognition capabilities with these universal models are increasingly appearing through firmware upgrades in products we already make use of in our everyday lives, as part of the emerging 'Internet of Emotions' (Pachalag and Malhotra, 2018). What's more, many products already analyse the emotions detectable not only in our faces but also in our voices. Digital voice assistants are a case in point: market-leading devices such as Amazon's Alexa, Apple's Siri and Google Home or Assistant are increasingly developing new services that are based on emotional cues picked up through vocal intonation, word choice and syntax. These cues are similarly premised on the idea that a series of core emotions can be effectively assessed and identified universally, across languages and cultures – even while many smart speakers are oriented towards a limited range of languages and markets. Alexa, for example, supported just eight languages (English, French, German, Hindi, Italian, Japanese, Portuguese-Brazilian and Spanish) as of October 2021.

In this chapter, we explore how the automated decision-making (ADM) carried out by voice assistants uses emotions as the basis for that decision-making and show how those decisions are tied up in services that are intended to automate different

DOI: 10.4324/9781003170884-8

forms of care that resemble 'women's intuition', or the feminised labours historically associated with high levels of emotional intelligence and gender (Hochschild, 2012; Sadowski et al., 2021; Strengers and Kennedy, 2020). We focus on Alexa and its associated 'Echo' devices, as the world's most widely used digital voice assistant run by the largest e-commerce company in the world: Amazon. Drawing on Alexa (and digital voice assistants more broadly), we describe a cloak of automated feminisation that conceals very human and deliberate decisions behind the programming of emotion recognition and emotion-based ADM. We contend that the femininity coded into these 'smart wives' is strategically positioned to facilitate the acceptance of caregiving being provided by such devices (Strengers and Kennedy, 2020).

We reveal four specific human decisions that underpin the ADM programming in these devices on (1) how to define and categorise emotions; (2) what data to collect on the users of these devices; (3) how these devices should look and sound; and (4) how care is understood and programmed. Through this discussion, we contribute to long-standing debates in anthropology, sociology, geography and science and technology studies (STS) that have critiqued the idea that we can universalise emotions and care (Lupton, 1998; Wajcman, 2017). In addition, we contribute to emerging critiques of voice assistants and other smart technologies that use their femininity to mask the extraction of data from people under the guise of maternal care (Bergen, 2016; Sadowski et al., 2021). We draw on our past research, particularly content analyses of popular media and trade press articles about digital voice assistants and automated technologies, promotional materials related to digital voice assistants and related devices, patents identifying the automation of emotion recognition and industry reports on affective computing (Strengers and Kennedy, 2020). For this chapter, we analysed these materials to identify and categorise the human decisions underpinning the developments of emotional ADM in voice assistants. We begin later by introducing the literature and critical debates that are informing these developments in the rapidly growing field of emotion-sensing technology and its emphasis on care. Our focus is on the automated emotional detections these devices can or are anticipated to make and the automated responses they provide in response to that detection. We focus on four human decisions that underpin these developments, before concluding with a warning about the need to explicitly call out and acknowledge the deliberate and conscious decisions that underpin ADM in the development of AI systems such as voice assistants.

Affective computing and the universality of emotions

Affective computing brings together computer science, psychology and cognitive science, aiming to design systems that can recognise, interpret and even mimic emotions to facilitate human–machine interactions. Such technologies contribute to reconfigurations of aspects of everyday life and to processes of socialisation. As advances in affective computing develop, they facilitate a more 'natural' human–machine interaction, making the presence of technologies even more immersive.

Affective computing endeavours to develop programmes and devices responsive to, or deliberately designed to influence, human desires and needs, as well as develop technology able to 'express' emotions. One of the central critiques of early affective computing, as pioneered by Rosalind Picard in 1995, was that it reproduced the forms of processing found in cognitivist applications by reducing affect to discrete units of information that can be captured, modelled and augmented (Boehner et al., 2005: 59). The social, interactional approach that has emerged in affective computing over recent decades is more culturally grounded and appreciative of how emotions are dynamically experienced. It focuses on identifying human emotions through signals such as facial expression recognition, facial action detection, gaze estimation, body language recognition and human physiological signal (heart rate) estimation.

The (ongoing) quest for affective computing is marked by a series of human and discipline-specific decisions that have led to the automated emotional detection we now take for granted in AI systems. The affective dimension of individuals is central to social life (Bendelow and Williams, 1998) and has been integral to Western approaches to studies of the human condition – for example, Descartes and Spinoza. Yet the so-called 'affective turn' in the early 2000s (e.g. see Gregg et al., 2010) placed emotions as the object of study across multiple disciplines including computer science, anthropology, media studies, economics and neuroscience. Each field emphasises the role of emotions in all aspects of human life, the latter for example, emphasising the role of emotions in brain function and processing.

There is little conceptual consensus across disciplines in terminologies of emotions or affect. Each discipline takes its own theoretical approach to emotions, which are typically explained from a neurobiological or sociological perspective. Emotions, from a cultural studies lens for example, can be described as 'socially and culturally codified feelings that can be shared or transferred between people, and that condition the way that humans relate to and participate in intimate, social, and political life' (Padios, 2017: 209). However, the most important theoretical approach in terms of this chapter, and of our own disciplinary orientation, is that found in the sociology of emotions (Stets and Turner, 2007) which posits that experiences and expressions of emotions cannot be understood outside of the social context in which they occur. Such perspectives stand in direct opposition to the theories of universal human emotions that are used in the programming of affective computing.

Examples of affective ADM technologies on the market include the 'Feel' wristband (www.myfeel.co/) and 'MoodMetric' ring (https://moodmetric.com/) which each uses sensors on the skin to read and respond in real time to electrodermal activity such as sweat, pulse and skin temperature. These physiological signals are indexed and measured, then parsed by algorithms to determine activation of the sympathetic nervous system which is then automated as emotional state feedback. These devices not only provide immediate state feedback displayed through a connected smartphone app but also parse the data into time-series data patterns. In these products, these data are fed back to the user through an app in which a variety of actions are automated in response, including more complex support functions such as mental health coaching. As Martin Berg (2017) writes, what data are collected, how algorithms interpret that data and how that data are translated in such devices is designed to offer a sense of highly individualised and personalised structure.

Digital voice assistants and Alexa's Hunch

In the growing market of digital voice assistants, there is considerable interest in using ADM to analyse vocal patterns and identify the emotional states of users. Amazon's digital home voice assistant 'Alexa' is one such company with a growing interest in these possibilities. Amazon's Alexa feature known as 'Hunches' aims to learn from users' patterns of behaviour to recommend or automate certain commands such as turning off the heater or lights when someone leaves the house or goes to bed. This, though, is just the beginning, with Hunches potentially able to pre-empt a user's moods, needs and desires through machine learning, in order to recommend appropriate products, solutions or courses of action.

The feature works by Alexa getting a 'hunch' about some aspect of your behaviour and making suggestions based on this. When the product was launched by Amazon, the company claimed that the feature aimed to replicate human curiosity and insight by programming intuition (Harris, 2018). Much like the other features of voice assistants which we have previously characterised as 'smart wives', Hunches is an arguably feminised skill that fulfils some of the many roles of the traditional housewife: to remember small details (like accidentally leaving the front light on) and help occupants realise small conveniences and pleasures (like turning on the heater before they get home). To date, Hunches are limited to automating other smart home devices and don't include the detection of emotions. However, patents involving the voice-detection of emotions lodged by Amazon in 2018 and other major voice assistant companies indicate they have bigger plans. For instance, Amazon's patent, 'voice-based determination of physical and emotional characteristics of users' (Amazon, 2018), allows Alexa to identify 'abnormal' bodily or emotional situations through voice: such as indications of coughs or sore throats, or excitable or sad behaviour such as laughing and crying. A voice processing algorithm applies tags to body and emotional attributes to establish baselines and monitor behaviour in an effort to detect 'happiness, joy, anger, sorrow, sadness, fear, disgust, boredom, stress or other emotional states'. Such cues, the patent explains, are based on 'an analysis of pitch, pulse, voicing, jittering and/or harmonicity of a user's voice, as determined from processing of the voice data'. Every user could have its own tailor-made emotional-detecting profile, identifying their 'default or normal/baseline state' in order to detect any 'abnormalities' and delivering tailored advice or content accordingly. Google also has a similar patent (dating back to 2014) to detect negative emotions in its users and to better assist them with whatever task is causing them unrest.

Of course, Amazon and other voice assistant developers are already using emotional detection in other ways, in an attempt to improve 'user experience'. For instance, in September 2019, Amazon launched a 'frustration mode', where Alexa will apologise if she detects a user becoming frustrated with ineffective request responses. It is clear from this emerging market and developments that there is a growing interest amongst voice assistant manufacturers to harness emotional detection to both enhance user interaction with voice assistants and also provide them with additional services. Where, though, have the (automated) decisions to provide these emerging services come from?

Very human decisions

In this section, we offer a critical perspective on how often unnoticed or seemingly unconscious automated responses and actions embedded into digital voice assistants impact the experience of emotion-sensing tech and its broader social ramifications. Focusing on four human decisions, we engage with important ethical debates that emerge from these decisions and draw attention to the implications for ADM and the human experiences they help mediate and facilitate.

Human decisions on emotional categorisation

Emotion-sensing technologies are designed, obviously, to detect emotions. As discussed earlier, emotional recognition detection through voice builds on the theories of emotional universality that have underpinned facial recognition systems for some time (Bryant and Barrett, 2008). There are a number of reasons for the very human decision to rely on these theories despite ongoing critiques and concerns about their ability to accurately detect emotional states. First, theories of universal emotions provide a practical conceptual framework that can be widely applied to emotion recognition systems through a small set of easily replicable principles (Crawford, 2021). Second, they open up the AI sector to a range of new and previously untapped markets. This is particularly important for a voice assistant system like Alexa, which is run by the largest e-commerce company in the world. Detection of emotional states can potentially lead to new products and services leading to currently under tapped markets, such as recommending psychology, health and wellness products and services to Alexa users. And third, the application of universal emotions – however clunky or lacking in nuance – serves to further 'humanise' feminised voice technologies such as Alexa, Google Home and Siri, by making them seem emotionally intuitive and empathetic and in turn increasing their likability and perceived trustworthiness (Strengers and Kennedy, 2020).

Humans make decisions on what specific emotions 'matter' and on what they mean in the context of the technology. As with the manual coding of Twitter data for sentiment analysis, the human work of deciding what represents which emotion, and what emotions to code for are highly selective and subjective decisions, where the privilege and intersectionality of the programmers have significant impact on what is selected and what is coded. Such practices of emotional extraction also make blind the subjective formation of emotions in the everyday lives of individuals insofar as they shape experience within social structures. It is well documented that the tech industry is heavily cisgender white male dominated and the AI and ADM sectors even more so. This lack of diversity translates to lack of understanding of diverse user needs (West et al., 2019a, 2019b).

The universality of understanding and application of emotions tend to pathologise human emotions and to reinforce existing regimes. Presuming that emotions are innate to all humans does not account for how people are socialised and trained to manage their emotions (Hochschild, 2012). Furthermore, when emotional intelligence is essentialised and presumed to be uniformly experienced and understood, diverse experiences and expressions of human emotionality are erased (Rhee, 2018). By replicating and reproducing this essentialist understanding of emotions into ADM, so too are experiences with that ADM viewed through this lens. Likewise, as found with facial recognition and interactions with robots, people may adapt their emotional reactions and responses in order to be correctly understood by the ADM, therefore, potentially modifying their expression of emotions to suit the detection capability of technology (Crawford, 2021; Rhee, 2018).

Fundamentally, AI is not nearly as capable of detecting emotions as developers claim and especially not when it is performing a task that requires human-level abilities (Barrett et al., 2019). A key concern is that facial expressions, or vocal cues, may provide little indication about a person's actual interior emotional world, as anyone who has 'put on a happy face' or tried to sound upbeat can attest. Such systems have also been found to have gender and racial biases in both facial and voice recognition, judging the speech affects of women differently from men, particularly Black women, and typically interpreting them as having more negative emotions such as anger (Crawford, 2021). Amazon's resume-screening AI infamously penalised female candidates. In 2018, CNBC reported that in response to video-screening being used to scan job candidates application videos for microexpressions and body language, candidates were advised to over-emote and/or wear additional makeup to make their faces more 'readable' to the AIs that screen the videos (Riley, 2018).

Despite these significant critiques, the ability of AI to map and automatically register not only facial features but also signifiers of emotions in our facial expressions, bodies and voice is part of the growing field of the Internet of Emotions, which aims to recognise and respond to human emotions through the use of ADM and also aims to automatically mimic human emotions through programmed voice and natural language processing. Importantly though, these interpretations and changes are not (only) the result of the automation but also of the human decisions that underpin its programming and detection capabilities.

Human decisions on data

Often decisions to collect data precede decisions on what to do with that data. Humans make decisions on what data to collect even when decisions on what forms of machine learning and ADM will be applied to that data have not yet been made. The imperative of data collection drives many business and tech development decisions. Emotion-sensing technologies that track movements and expressions reveal detailed information which is used to fuel advertising, product recommendations and profit. As quoted by Jathan Sadowski, Andrew Ng, a top AI researcher for the largest corporations including Google exposed the primacy of data collection: 'At large companies, sometimes we launch products not for the revenue, but for the data. We actually do that quite often . . . and we monetize the data through a different product' (Ng cited in Sadowski, 2020: 30).

In the case of emotion-sensing technologies, there are multiple processes of extraction occurring. Jan Padios defines how emotional extraction is purposefully operationalised in two distinct processes (2017). The first is the form of emotion resource transfer that occurs in the work of caring for others, but Padios also identifies this transfer in the production of new technologies such as emotionally aware devices because of the way they transfer emotional resources from one party/actor to another, for example from user to corporation. The second form of emotion resource extraction regards the use of emotion knowledge for the purpose of extrapolating human behaviours which is what data banks can be used for. This is described as occurring in a range of domains including service work, social media and AI where theories of emotions are used to understand data mined through the process of defining and recording people's emotions. Padios points out that emotion extraction processes are integral to understanding the intersection of culture and capitalism in which the drive for value and certainty are mobilised through surveillance and control (2017: 207–8).

Such concerns are illustrated through another member of the Alexa family – the Echo Look – a fashion assistant launched in April 2017 and discontinued in July 2021. We discuss the device briefly here in terms of data decision-making but social practices of decision-making for users of the devices are explored in greater detail by Heather A. Horst and Sheba Mohammid in Chapter 6 of this collection.

The Echo Look was similar to other Echo smart home devices. It operated Alexa, and could, for instance, control other devices in the home, play music or answer questions about your day. However, it came with one additional feature unique to other Echo products, an integrated camera. Using the camera, users were encouraged to take photos of their daily outfits to track and explore their fashion style. These photos were displayed back to users, augmented with fashion advice, and shopping experiences through virtual reality.

It is clear that a great deal of information can be gained, from the health of users to their mental state, or economic status. The data could also be used to identify lifestyle changes, including pregnancy. There are telling examples already out there, Facebook have stated they can potentially identity for advertisers the mental wellbeing of teens in real time through selfies (Davidson, 2017). Yet human decisions on how machine learning will be applied to such databanks are typically not made known to users.

Humans also make decisions in lieu of, or as proxies to, anticipated machine learning and ADM processes. Often a team of humans replicates what programmed and ADM or AI will supposedly do once fully built and implemented. This is not necessarily made clear to users. This is especially the case in the early testing phase of technology development and may only be a temporary measure, demonstrating what the AI is intended to do. Some technologies use humans to back up the

ADM. In this hybrid approach, the AI performs basic functions and humans are there to bridge current gaps in development. AI has the potential to scale large volumes of data, but for start-ups and companies with limited access to such volumes of data, it is more cost-effective to use humans rather than build AI. One of the issues with these approaches is that humans can manage far more complex and nuanced data than may be possible with the AI they are 'impersonating'.

Also mostly invisible through the imaginary of AI and ADM which supposes unsupervised machine learning are the labours of humans listening to the snippets of conversations and commands captured by smart home devices, transcribing and coding the content to better train the systems that augment understandings and applications of emotions hardwired into devices such as Alexa. Tuukka Lehtiniemi and Minna Ruckenstein discuss these 'hidden faces of automation' (Irani, 2016) in Chapter 12 of this volume, in which they describe the labour of Finnish prisoners labelling language data for a local AI firm. A 2019 report in *Bloomberg Businessweek* exposed transcription farming practices of big tech firms including Amazon, Google, Facebook, Microsoft and Apple, where contractors perform the harvesting and analysis of people's voices for tone and content, and background sounds that might provide other clues to people's lived experiences, such as children crying (Carr et al., 2019). This is an example of large-scale supervised machine learning, which involves humans labelling objects which are then used to train the machine, whereas unsupervised vision leaves the machine to identify categories/classifications based on latent data.

Hybrid approaches also impact user expectations of AI, with users expecting more sophisticated results from the technologies than they are capable of performing. ADM does best with very narrowly defined constraints, for example identifying a limited selection of emotions based on the presence of exaggerated expressions or tones of voice. The hybrid approach is sometimes embraced as an ongoing solution to the balancing of AI efficiency and human flexibility. When AI cannot recognise an object in an image, the query is sent to humans to categorise. Hybrid chatbots transfer conversations to humans once the conversation gets too tricky for its programming. This strategy has its own pitfalls. Users often do not respond well to discovering they have been talking to a human when they presumed they were disclosing personal information to a chatbot.

Human decisions on how users interact with devices and how devices respond

There are many human decisions in the designing and building of emotion-sensing technologies. For physical objects, these include decisions about the shape, size, texture and colour of materials to use. These also include decisions about what features to include, and how users interact with the device – is there haptic, sound or visual feedback, are there buttons or controls? In short, everything about the form and function of a device has been selected – very carefully – by a human or a team of humans. The devices we have available to us are very much limited by their human programming, even if much of this programming is intended to be invisible to the user.

We have written elsewhere on how the design of caregiving and assistive technologies are typically feminised, where voice assistants and social robots are given feminine voices, names, physical features, personalities and domestic purposes (Strengers and Kennedy, 2020). Studies of the industries which produce such technologies show significant gender disparity, with men vastly outnumbering women in roles in the fields of automation. This lack of diversity has impacted the types of devices being designed and helped fuel the development of smart wife technologies.

There is a significant issue of lack of diversity in the tech industry more broadly. Not only are men over-represented in the industry but so also are white people (West et al., 2019a, 2019b). The industry doesn't just reflect social inequalities, it aggravates them because the lived experience of those in the industry comes to influence what is designed and for whom. The tech sector is to some degree finally recognising issues of diversity and making efforts to broaden the ideas incorporated as core features of technologies being designed for everyday use.

While some human decisions are very visible (or audible), the presence of humans involved in the design of technologies are intentionally invisible. Never in Alexa's responses will you hear an admission that her feminisation, incompatibility, inoperability or functional limitations are a consequence of programming and very-human decision-making. Alexa willingly takes the blame for human inefficiencies and oversights. For instance, she will apologise repeatedly on demand whenever requested to do so, without even querying what it is she is apologising for (Strengers, 2021). Despite this response being automated, it is nonetheless a programmed one, and, therefore, one which a human at some point made, presumably without being aware of the potentially harmful gendered message this may send about the role of women in society. Indeed, the human decisions central to Alexa and other voice assistants' programming are so invisible that it is often the voice assistants' themselves (and their feminised personalities) that are blamed when the devices malfunction or are unable to answer or assist with a query. As we have described elsewhere, in such situations a voice assistant is often treated and discussed as a 'bitch with a glitch', with journalists and users frequently belittling and abusing feminised devices, rather than more accurately and appropriately directing blame and responsibility at the male-dominated teams who program them (Strengers and Kennedy, 2020). In these ways, the role of ADM in hiding the human decisions and the people who program digital voice assistants, reproduces and reinforces problematic gendered stereotypes about women in society.

Human decisions about care

The human decision to rely on universal indicators of human emotions through specific facial or vocal cues is also redefining the meaning of care in relation to our interactions with emerging AI. Following Judy Wajcman (2017: 123), we are concerned that features such as Hunches 'mistake the appearance of care with real empathy and genuine personal interaction'. This is particularly the case given the feminisation of caregiving programmed into voice assistants, as discussed in the previous section. Deploying feminised and wifely stereotypes, Alexa's attempts to pre-empt a user's wants and desires via Hunches, and at some point in the future also respond to their emotional state, situates care as replicable and programmable feminine traits that can be addressed through individualised and commodified products or services. Such approaches position caregiving as a form of emotional extraction and responsiveness, which Padios (2017) describes as occurring within paradigms of control and commodification which pay little attention to diversity in cultural and social application. Such emotional extrapolation is likely to be unevenly experienced by vulnerable groups, including those who are non-white, non-male or those who have forms of neuro-diversity.

Aside from potentially failing to account for the diversity of emotional representation and the attuned caregiving that is commonly required to respond to it, there is something even more worrying about this new interpretation and delivery of caregiving through vocalised AI: it's ability to redirect and redefine how we come to understand and practice caregiving, and intimacy more broadly. Through ongoing, uniform and mass distributed forms of emotional detection and caregiving actions, the human decision behind emotional programming is redefining the caring landscape. This can be seen not only in the ambitions of device manufacturers themselves as revealed through products or patents such as those outlined in this chapter but also in nation-state and corporate mission statements, such as political commitments to social and affective robots as a 'solution' to eldercare and aging populations (Robertson, 2010). As Wajcman points out, when caregiving gets redefined as what can be delivered by and through affective devices and robots, other forms of caregiving services and models are overlooked. These include revaluing and remunerating care labour the same as we remunerate programming work, addressing caring labour shortages, or as Wajcman (2017: 123) more radically proposes, redesigning housing and cities 'so that the elderly were not relegated to separate places but were integrated into the wider civil society'.

In the case of Alexa's Hunches, intentions to provide emotional care can be interpreted as an attempt to commodify and capitalise on emotional labour. 'Hunches' is in itself a clever marketing term, resembling the traditional and elusive notion of 'women's intuition', or a very human form of emotional intelligence. Feminised through its default voice and name, Alexa's 'Hunches' also reinforces the idea that women are the primary caregivers gifted with 'natural' emotional intelligence, and that they have endless reserves of understanding, while lacking any emotional needs of their own. As illustrated through Alexa's 'frustration mode' described earlier, and through the device's ability and willingness to apologise on demand (Strengers, 2021), the programming priority for this form of 'care' is to appease users at all costs and even at the expense of the devices own 'self-care'. While of course devices *themselves* don't need care in the sense we are discussing here, we and others have argued elsewhere how humanised devices reinforce gendered stereotypes and can perpetuate sexism and violence towards women through the decisions that humans program into them (Strengers and Kennedy, 2020). In addition, and as Hilary Bergen (2016: 107) has argued, attempts that allow AI to respond to human emotional states are 'a development that poses a severe threat to human rights and privacy. What may seem like empathy is really an act of manipulation'.

Finally, there is a further implicit devaluing of feminised labour present in this form of programmed caregiving; namely, the assumption that caring and emotional labours *can* and *should* be outsourced to a device or robotic deputy such as Alexa. Such an assumption can itself be viewed as a form of undermining and simplifying the highly complex roles and tasks that women traditionally and still predominantly perform as society's default nurturers, carers and emotional labourers.

While many supposedly emotionally intelligent devices and developments are well intentioned and potentially beneficial for people experiencing mental health issues or physical conditions that require forms of emotional care, we also have to ask what these human decisions that are reproduced through ADM cost us as they continue to redefine what caregiving is, who can receive it, and how and who should perform it.

Conclusion

In this chapter, we have highlighted how the development of emotion-sensing technologies – and ADM more broadly – are predicated on very human decisions that lead to a series of problematic outcomes. The four types of human decisions we have highlighted are complicated by presumed universality of emotional representations, profit-driven data extraction, over-estimation of automated decision-making and AI capabilities, lack of diversity in the industry sector and the decision to represent devices as representations of women and equivicising of caregiving labour with empathetic caring. Each of these decisions has a series of potential consequences and problematic impacts which we have highlighted in this chapter, ranging from the reduction of emotional expression and interpretation through to the devaluing of caring labours in society.

ADM's role in facilitating these impacts is important and often overlooked. By its very characteristics, ADM gives rise to the perception that humans are no longer involved in the responses, interpretations or outcomes of its applications within a technology such as digital voice assistants. This of course is not the case, but this illusion of automation is where the most problematic aspects of ADM lie, and where researchers, designers and programmers must further focus our demystifying efforts. As we and others have made clear, people and human decisions are absolutely central to ADM, and it is only through the exposure and ongoing accountability of these human decisions that we will be able to ensure that ADM can develop ethically and responsibly (if indeed that is still possible). By busting the myth that ADM is only at fault through unconscious bias or design flaws, we can begin to expose, demand accountability for, and redress the human decisions that fuel its existence. As our example of Alexa's Hunches and digital voice assistants has shown, the decisions behind ADM are not invisible and accidental but rather intentional and calculated. They deserve our exposure and scrutiny to ensure that they develop in a way that is beneficial to society and advance the gender and other forms of progress many have fought so hard to achieve.

References

Amazon (2018) Voice-based Determination of Physical and Emotional Characteristics of Users. Patent No. US10096319B1. Available at: https://patents.google.com/patent/US10096319B1/en

Barrett LF, Adolphs R, Marsella S, Martinez AM and Pollak SD (2019) Emotional Expressions Reconsidered: Challenges to Inferring Emotion From Human Facial Movements. *Psychological Science in the Public Interest* 20(1): 1–68.

Bendelow G and Williams SJ (1998) *Emotions in Social Life: Critical Themes and Contemporary Issues*. London: Routledge.

Berg M (2017) Making Sense With Sensors: Self-Tracking and the Temporalities of Wellbeing. *Digital Health* 3(1): 1–11.

Bergen H (2016) 'I'd Blush If I Could': Digital Assistants, Disembodied Cyborgs and the Problem of Gender. *Word and Text, a Journal of Literary Studies and Linguistics* 6(1): 95–113.

Boehner K, DePaula R, Dourish P, et al. (2005) Affect: From Information to Interaction. In: *Proceedings of the 4th Decennial Conference on Critical Computing: Between Sense and Sensibility*, 20 August 2005. CC'05. New York: Association for Computing Machinery, 59–68. https://doi.org/10.1145/1094562.1094570

Bryant G and Barrett HC (2008) Vocal Emotion Recognition Across Disparate Cultures. *Journal of Cognition and Culture* 8(1–2): 135–48.

Carr A, Day M, Frier S and Gurman M (2019) Silicon Valley Is Listening to Your Most Intimate Moments. *Bloomberg Businessweek*, 11 December. Available at: www.bloomberg.com/news/features/2019-12-11/silicon-valley-got-millions-to-let-siri-and-alexa-listen-in

Crawford K (2021) *The Atlas of AI*. New Haven, CT: Yale University Press.

Davidson D (2017) 1 May. Facebook Targets 'insecure' Young People. *The Australian*. Available at: www.theaustralian.com.au/business/media/facebook-targets-insecure-young-people-to-sell-ads/news-story/a89949ad016eee7d7a61c3c30c909fa6

Gregg M, Seigworth GJ and Ahmed S (eds) (2010) *The Affect Theory Reader*. Durham, NC: Duke University Press.

Harris M (2018) Amazon's Alexa Knows What You Forgot and Can Guess What You're Thinking. *The Guardian*, 21 September. Available at: www.theguardian.com/technology/2018/sep/20/alexa-amazon-hunches-artificial-intelligence

Heaven D (2020) Why Faces Don't Always Tell the Truth About Feelings. *Nature*, 26 February. Available at: www.nature.com/articles/d41586-020-00507-5

Hochschild AR (2012) *The Managed Heart: Commercialization of Human Feeling*. Stanford, CA: University of California Press.

Irani L (2016) The Hidden Faces of Automation. *XRDS: Crossroads, The ACM Magazine for Students* 23(2): 34–7. https://doi.org/10.1145/3014390

Lupton D (1998) *The Emotional Self: A Sociocultural Exploration*. London: SAGE.

Pachalag V and Malhotra A (2018) Internet of Emotions: Emotion Management Using Affective Computing. In: Satapathy S and Joshi A (eds) *Information and Communication Technology for Intelligent Systems (ICTIS 2017) – Volume 2*. Smart Innovation, Systems and Technologies. Cham: Springer, 567–78. https://doi.org/10.1007/978-3-319-63645-0_63

Padios JM (2017) Mining the Mind: Emotional Extraction, Productivity, and Predictability in the Twenty-First Century. *Cultural Studies* 31(2–3): 205–31.

Rhee J (2018) *The Robotic Imaginary: The Human and the Price of Dehumanized Labor*. Minneapolis, MN: University of Minnesota Press.

Riley T (2018) Get Ready, This Year Your Next Job Interview May Be With an A.I. Robot. *CNBC*, 13 March. Available at: www.cnbc.com/2018/03/13/ai-job-recruiting-tools-offered-by-hirevue-mya-other-start-ups.html

Roberts K, Roach MA, Johnson J, et al. (2012) EmpaTweet: Annotating and Detecting Emotions on Twitter. In: *Proceedings of the Eighth International Conference on Language Resources and Evaluation (LREC'12)*, Istanbul, Turkey, May, 3806–13.

Robertson J (2010) Gendering Humanoid Robots: Robo-sexism in Japan. *Body & Society* 16(2): 1–36. https://doi.org/10.1177/1357034x10364767

Sadowski J (2020) *Too Smart: How Digital Capitalism Is Extracting Data, Controlling Our Lives, and Taking over the World*. Cambridge: MIT Press.

Sadowski J, Strengers Y and Kennedy J (2021) More Work for Big Mother: Revaluing Care and Control in Smart Homes. *Environment and Planning A: Economy and Space*. https://doi.org/10.1177/0308518X211022366

Stets JE and Turner JH (eds) (2007) *Handbook of the Sociology of Emotions*. Berlin; Heidelberg: Springer.

Strengers Y (2021) Amazon Echo's Alexa Is Programmed to Always Apologize – Especially When It's Not Her Fault. *Think*, 2 March. Available at: www.nbcnews.com/think/opinion/amazon-echo-s-alexa-programmed-always-apologize-especially-when-it-ncna1259001

Strengers Y and Kennedy J (2020) *The Smart Wife: Why Siri, Alexa and Other Smart Home Devices Need a Feminist Reboot*. Cambridge: MIT Press.

Wajcman J (2017) Automation: Is It Really Different This Time? *British Journal of Sociology* 68(1): 119–27. https://doi.org/10.1111/1468-4446.12239

West SM, Kraut R and Chew H (2019a) *I'd Blush If I Could: Closing Gender Divides in Digital Skills Through Education*. UNESCO and EQUALS Global Partnership. Available at: https://unesdoc.unesco.org/ark:/48223/pf0000367416

West SM, Whittaker M and Crawford K (2019b) Discriminating Systems: Gender, Race and Power in AI. AI Now Institute. Available at: https://ainowinstitute.org/ discriminatingsystems.html

6

FRAMING FASHION

Human–machine learning and the Amazon Echo Look

Heather A. Horst and Sheba Mohammid

Introduction

Over the past decade, new automation technologies have become part of our everyday lives, influencing the kinds of decisions that people, organisations and institutions make. Data scientists, computer scientists and others involved in the design, development and dissemination of technologies seek to enhance interactions between humans, human data and machines by training machines to replicate human decision-making 'by exposing them to a large number of examples and rewarding them for drawing appropriate distinctions and making correct decisions', much in the same way as human beings learn (Lazarus et al., 2018: 6). Yet, as the extensive work on learning and pedagogy demonstrates (Cole and Engestrom, 1989; Lave, 1988), human forms of learning are not simply cognitive. Learning is embodied, rooted in culture, situated in social contexts and shaped by ideology, power and social practices (Cole and Engestrom, 1989; Freire, 1970; Lave, 1988). As Vygotsky and Cole (1981) suggest, learning is 'culturally shaped by the social environment in which it takes place' (Smagorinsky, 1995: 93) resulting in a learning dynamic wherein 'culture and cognition cocreate one another' (Cole, 1985: 3). Different learning styles as well as different modalities of teaching and creating knowledge also influence the learning process (Gardner, 2000). While cognitive processes integrated into machine learning may be the beginning of automated decision-making (ADM), more complex forms of artificial intelligence (AI) require enhancements in the dialogical approaches in order to more effectively take into account unforeseen situations and scenarios to reflect the ways in which humans learn.

This chapter examines learning dynamics between humans, human data and machines through the study of a new, consumer example of machine learning and AI designed to help people make decisions about what to wear, The Amazon Echo Look (henceforth Echo Look). Integrating information from sites such as Instagram,

DOI: 10.4324/9781003170884-9

where people posted and reviewed clothing and input from professional stylists, the Echo Look used machine learning and AI to provide people with feedback on different clothing options. It also promised to learn from the participants about their preferences over time to provide more customised advice. Drawing upon a study of 25 women in the USA and Trinidad, we explore the contexts and content that our participants reported the device overlooked and the consequences of these gaps for the ways in which our participants perceived the possibilities of the Echo Look. Our chapter responds to a call to develop nuanced understandings of the human–machine interactions and the potential for more expansive forms of ADM, from the vantage point of the people using these new applications (Rahwan and Simari, 2009).

Closet ethnography and the Amazon Echo Look

Avery, a 30-year-old, Japanese-American woman living in San Francisco, California, started using the Amazon Echo Look in early 2020. Rather than consulting the instruction manual, Avery began the process of learning about the device by placing the Echo Look in her bedroom near her closet. As someone who pursued photography as a hobby, she liked the pictures the device took and admired the effective use of the Bokeh technique which involved blurred out the background and while leaving the participant in focus. She also relished the fact that she could see her entire outfit, a feature that enabled Avery to pause and reflect on the ensemble. This moment allowed Avery to contrast the way the outfit looked in her imagination and the reality of how this particular ensemble looked on her own body. The ability to see the way an outfit 'really' looked surpassed the use of the mirror and, from her vantage point, improved her decision-making about what to wear.

Sun and Zhao (2018) assert that disruptive advancements in digital technology ranging from AI, robotics, additive manufacturing and radio frequency identification to the virtual dressing room, e-commerce, social media and other emerging digital integrations are becoming key drivers in the fashion and lifestyle industries (Knight, 2017). While the application of AI to fashion has occupied the imagination for decades (Luce, 2018), they are only now becoming a technical reality and available on the consumer market. The Echo Look is one such example. Released in 2017, the Echo Look brought together data from sites such as Instagram where people post and review clothing, input and feedback from professional stylists and information from device users, to provide recommendations for clothing. Built with machine learning and AI technologies, it promised to learn from the participants about their preferences to provide more customised recommendations over time. The device was first introduced to a small specialist demographic of fashion enthusiasts in 2017 and was later broadened to include wider consumers in the USA. Although marketing was limited, it was offered on Amazon at a cost of US$199.00. With the support of the device's voice-activated camera, users could stand in front of the camera's line of sight and try on outfits; when they were changed and ready, they asked Alexa to take a picture. The Echo Look used a depth-sensing camera, LED lighting and computer vision to blur the background of an individual's shots.

These shots and videos were integrated into the app so users could archive, review and categorise their wardrobe and compare outfits using the Style Check feature. These were then compared and ranked by the Style Check with a percentage preference and a descriptive rationale for the ranking.

The Echo Look was discontinued in June 2020 with industry reports noting that many of the Echo Look's features were already found in the shopping app. Others speculated Amazon would benefit from selling its products via recommendations to a larger group of users and its integration into other Amazon applications would enable more widespread access to the features (McGlaun, 2020). Indeed, Echo Look features were eventually absorbed into the Amazon App and the Amazon Echo Show device, and offered as a partial replacement as the Echo Look to consumers who purchased the device. Because Amazon recommended the device be set up in closets and bedrooms, tech bloggers noted the problem of having a camera that could photograph one while in various states of undress also contributed to its small uptake (Surur, 2020). These challenges mirror the broader academic critique of the device and application in relation to issues of privacy and the corporate ownership of data (Barrett, 2017; Ramadan, 2019; Strengers and Kennedy, 2020; Vincent, 2017).

Given the capacities of the device, its use in the home and the prevalence of Amazon accounts in the USA and the Caribbean, we developed the closet ethnography as a method for pinpointing how the Echo Look might shape decision-making about what to wear. In anticipation of Amazon's broader promotion of the device in the future, we carried out the small study with 25 participants in the USA and Trinidad in late 2019 and the first half of 2020. All of the participants were women between the ages of 30 and 50 years, the target age demographic for the device, and were recruited through the authors' personal and professional networks. Building upon ethnographic studies of clothing and wardrobes (Woodward, 2007), we sought to document the suite of clothing in our participant's closets, drawers and wardrobes and their experience and perspectives of using technologies.

We began the study by providing participants with a new Amazon Echo Look device to use over the course of the study. Sheba Mohammid either went in person to install and assist with setting up the device or talked the participant through the process over the WhatsApp platform. We conducted two sets of semi-structured interviews with participants. They were interviewed before they used the device so that we could get to know them better and gain a sense of their interest in and general attitudes towards style, fashion and technology. This included understanding participants' fashion influences, changes and aspirations as well as their dressing and shopping habits and routines. These interviews also framed an inquiry into the meanings participants attached to fashion and their views on the socioeconomic significance ascribed to styles in their regions.

The closet ethnography continued with a diary study where participants used the device for three consecutive days over a three-day period, in keeping with the original marketing of a device as one that could be made for everyday decisions about what to wear. Afterwards, we asked participants to share or capture images/screenshots of these outfits and the recommendations for discussion during the follow-up interview. However, we found that most of the participants wanted to keep the device for at least two weeks to gain better familiarity with the technology. In addition, they chose to batch their Echo Look use into an overview of their wardrobe. Instead of a short instance of daily use to decide between a couple of outfits while getting dressed, participants preferred to devote an entire afternoon or evening usually on a weekend when they had time and tried on many different outfits. They felt that they did not have time in the morning to switch outfits or make key decisions given all the other things they needed to do in their lives (e.g. children and commuting). Other participants approached their outfit not as a choice between two outfits when getting ready but instead noted that the 'getting ready' process started in advance. Some people planned their outfits at the beginning of the week and others preferred to make choices based upon previous experience of what was comfortable or well received. In essence, they asserted more of a conceptual understanding of their wardrobe, the constitution of an outfit and the other factors that shaped decisions around style and what to wear. A formal semi-structured interview and follow-up discussions for up to six months were also conducted with participants.

What the Echo Look doesn't see

In a popular CNN interview on the future of getting dressed, Kavita Bala noted that one of the main challenges of using AI and machine learning technologies in the context of fashion is tied to the perception that human style can be distinctive, subtle and individualist (Yurieff, 2017: 29). In the following section, we introduce the wardrobes of four women who participated in our study, with a particular focus upon how they began to integrate the use of the Echo Look into their routines. We discuss the ways in which they imagined the Echo Look might help with their daily dilemma: the question of what to wear. In addition to illustrations of its use and potential, we also highlight the ways in which the device sometimes fell short of their expectations. Specifically, we focus upon two areas – content and context – where our participants did not feel the Echo Look provided them with the experience of recommendation and feedback that they anticipated.

Fashion content

This section begins with a return to Avery who relished the opportunity to obtain a 360 degree view of her outfits. Alongside the camera view, Avery noted that one of her favourite features was the Style Check which provided feedback on outfit options. Avery enjoyed comparing the recommendations with those from her partner whose advice sometimes differed from the suggestions provided by the device. Avery noted that she often had to reject the recommendations of the Style Check because it was missing some information. For example, one key parameter that Avery felt was important in her choice of attire was comfort. During the COVID-19 pandemic, there was a period during which San Francisco was issued a 'Shelter-In-Place' order with non-essential workers like herself (who worked in HR) required to remain indoors. For this period, Avery spent much of her time at home on Zoom calls and even when she was able to go into the office, she pursued a sense of ease in her dressing during her day. She felt that her lifestyle changes and emotional state during this time motivated her to seek comfort even more as a factor in choosing an outfit than she might have done in the past. She likened this to a metaphor of a chameleon merging with the background where she felt she was engulfed in the energy of the outfit; just knowing she was in a certain colour made her feel better. Avery described how dressing, at least for her, was analogous to knights wearing armour, where what they wore could affect how they functioned and how secure they felt. Avery's criterion for comfort was connected to a particular piece of clothing's fit, feel and texture, which were not aspects of the clothing that the device gave recommendations about. Nor did the device realise her new-found propensity to seek more cheerful colours to combat the bleakness she felt accompanied the pandemic.

Kenna, a 28-year-old, Afro-Trinidadian dentist who lived near the Trinidad capital, Port of Spain, participated in our study for two weeks in late 2019. Kenna had an expansive wardrobe and her closet was meticulously organised. Once she received an Echo Look, she began taking pictures of her entire wardrobe and categorised her outfits into compartmentalised zones of her closet with gym 'looks', going out 'looks', carnival 'looks', beach 'looks' and work 'looks'. She found this taxonomy useful when using the Style Check feature of Echo Look and she used this to compare outfits that she would wear to the same context and social space. She was impressed by the Echo Look's ability to discern types of outfits, as the app created a category labelled 'business attire' on its own and placed her work looks into this section. Kenna also enjoyed the ability to see herself in a 360 degree view and noted that in the process of trying on outfits the device enabled her to see that a few of her skirts were a little too short in the back: she moved these out of her regular rotation.

Like Avery, Kenna was very keen to try the Style Check. But after using it for a while, she realised that there was also a tension between what she liked and wanted to wear and what the device recommended. As a modern Trinidadian woman, Kenna was very keen on wearing pants for a series of practical and aesthetic reasons. But every time she compared outfits, the device consistently ranked work outfits that contained skirts rather than trousers more favourably. Kenna noted that the app said that the outfit shapes looked better with the skirts and she wondered if she 'lost her shape' in pants. Kenna questioned if the stylists and the algorithm that were used to create the Echo Look were predisposed to certain traditional norms of an appropriate outfit for a woman to wear to work. She felt a distinct unease with what she perceived as a gendered idea of how a professional woman should look. As she described, 'I mean it kind of made me feel like, "Why are you trying to tell me what to do? Why are you trying to tell me as a woman that I can't wear pants to work?"'.

Alongside style or cut, Kenna also found that the app demonstrated an aversion to bright colours. 'Better colour combinations' in the Style Check preference feedback always meant black and white combinations or muted, monochromatic clothing. These did not reflect the style in urban Trinidad and made Kenna wonder if the device not only missed the mark in terms of local style but also could be missing out on trends. Kenna mused that in some fields there are people who often stick to one sort of colour palette, but she believed that she had the freedom to experiment with colour in creating her work outfits and valued this creativity. Kenna went as far as to say that she felt the device 'judged' her choices and provided an antiquated perspective of what constitutes professionalism. In the end, and despite having 'fun' experimenting with the Echo Look, Kenna felt there were too many values inscribed into the device to be of reliable use in her everyday life.

Fashion contexts

If the types of appropriate professional clothing recommended by the Echo Look missed the mark for Avery and Kenna, other participants felt the device missed out on the contexts in which clothing might be worn. For example, Jacqui was a 55-year-old woman originally from Trinidad and Tobago who lived in Orange County California, USA, who participated in our study in late 2019. Of mixed Spanish and Asian heritage, Jacqui noted that her Caribbean origins gave her an appreciation of bright colours but that she was always careful to conform to norms in the USA. Jacqui described how her family and friends in Trinidad devoted lots of time to fashion and were often more critical of her appearance than her American friends and colleagues. Jacqui considered herself quite tech-savvy and was excited to use the Echo Look which she thought might shed light on 'where we had come' with technology. Jacqui used the Style Check feature of Echo Look to compare outfits and found that colour and fit seemed to be the most common determining factors behind recommendations. While Jacqui had always been interested in choosing colours that she felt flattered her in pictures, she found it difficult to agree or disagree with feedback on an outfit devoid of context in which it would be worn. As she assessed the suggestions of Echo Look preferences, Jacqui noted that the success of an outfit for her largely relied not only on its fit on her body but also on its appropriateness for the occasion in which the outfit would be situated. This included perceived levels of formality, weather and other environmental conditions and, most often, what others would be wearing.

Like Avery, Jacqui also described how being able to see the outfits side by side was useful in giving her a wider perspective. Yet, whether or not she would actually wear an outfit depended on the context in which it would be worn. As she described,

> you are able to also be more critical and think about the suitability of the outfit if you think of where you are going . . . because I can also say it's a little bit too overdressed in the context of where I'm going and knowing how the other people are dressed. (Echo Look) may pick this (outfit), yes, because of these factors, but this (other outfit) may work more . . . in the Trinidad environment. . . . So you have to think back to where you're going and the people you're with.

When reviewing Echo Look's suggestions, Jacqui was conscious of not being overdressed, which she felt might alienate her from her neighbours in suburban Orange

County. She felt that while an outfit suggested by Echo Look might suit her or be quite stylish, she may still not opt to wear it because it felt too dressy for California. She even took subtle reconfigurations of an outfit into account. For example, Echo Look preferred a tied-up version of a top more than a looser one, suggesting that it was more flattering on her because of fit. While Jacqui could see why this style suited her silhouette, she concluded that she would never wear that look in California with her friends 'because it looks a little more dressy than what they tend to wear. . . . Context is very important'.

Our final example comes from Quinn, an Anglo-American mixed-media artist in her early 30s who lived in San Francisco, California, and participated in our pilot study during the first half of 2020. Quinn talked a lot about how she enjoyed playing with the device and trying different versions of outfits. Sometimes she tried tucking in her shirt or reconfiguring the same outfit to see what option the device recommended. Quinn often found that the Echo Look made interesting and useful suggestions that provided her with a different perspective on what she might wear. Being able to see her body fully and how an outfit looked on her body in various poses also replaced the use of a mirror. Quinn also appreciated the ability to see how the same outfit may compare quickly and she found the data that the Echo Look provided on preferences on colour and fit to be informative. As she characterised it:

> I guess it was a little bit like receiving an outside perspective instead of thinking in my head how I looked . . . you're getting how you present to yourself, not how you present to the world. . . . It's like a breath of fresh air because it's a vision of what I look like from the outside.

While Quinn found the Echo Look useful, she still positioned its feedback as part of a range of variables that went into the decision of what to wear. She thought that the decision-making process on outfit choice could be enhanced through more dialogue with the Echo Look, including questions regarding where the outfit was being worn.

Like Jacqui, Quinn found that it was crucial to situate the Echo Look's feedback within the social and physical setting of where she was going. There were instances where she was ambivalent towards the outfit Echo Look preferred or opted to wear a different one as it 'was more appropriate for that place'. In fact, Quinn thought the Echo Look experience would be improved by learning more about the participant and collecting data to better inform its preferences and suggestions including:

> the different contexts for a look analysis. Stuff like, "Are you going to a party or are you going to work? Are you walking around the neighbourhood?" whatever it is. I think the context would be a little helpful.

Quinn also shared the view of other participants that the Echo Look could use data that may be available to the device such as weather conditions in order to provide more accurate recommendations on what to wear, including needing a jacket, pullover or a sweater. Participants who were familiar with Amazon products expressed surprise that these features were not already integrated into the device. Quinn, for example, lamented the many times she was caught without a jacket or her shoes turned out to be inappropriate for weather changes on her commute back from work. If the Echo Look consolidated this information into the recommendations, it could be a valuable feature that made the most of other information that the device may have access to, without the need to search through multiple sources.

Dialogical learning and the Amazon Echo Look

Determinations about fashion and what to wear involve a range of factors. Loschek (2009) argues that fashion involves reciprocity and interaction between the designer, object and viewer and individual determinations of being 'well dressed' requires identifying clothes that suit the individual's body and their sense of style defined by the individual and the larger world in which they exist (Norell et al., 1967). In many ways, this dialogical sense of fashion is intertwined with the ways in which our participants approached their use of the Echo Look. They viewed the Echo Look as a 'beta' product, positioning it in a spectrum of developments in technology and fashion, and they were interested in the ways in which their use of the device might shape future iterations of it. It also meant that, at the outset, they framed it as a device through which they (and the device) might learn together. Participants were often less interested in the exact outfit chosen by the Style Check feature and more intrigued by the logic that underpinned the recommended outfit. They used the specific feedback given by the device to build a base of knowledge and repertoire of skills in selecting what types of outfits looked best on them in terms of characteristics such as colour, style and fit. Most participants felt that the device was useful in helping to provide the expertise for them to gain skills using an outside perspective on the types of clothes that might best suit their bodies and colouring. Yet, without an understanding of the contexts in which the outfits would be worn, the value of the expertise promised by the device was compromised.

The receptivity of participants to the Echo Look and the possibilities for dialogical interaction with the device for learning about fashion and style appeared to deliver upon the aspirations of those who design, build and market technologies based upon machine learning and AI. Yet, the gaps in 'seeing' fashion content and context ultimately demonstrated that the machine–human interface of the device resembled much more of what we might think of as a 'banking' model of learning, to follow Freire's (1970) metaphor of learning and pedagogy. A 'banking model' treats teaching as an act of depositing information which is passively received by a student who is expected to memorise and repeat information to demonstrate learning (Freire, 1970). By contrast, a dialogical approach to learning involves the creation of a space for students and teachers who teach and learn from each other (Shor and Freire, 1987). Students become active agents of critical inquiry where they join with the teachers in 'reading' the world and teachers also learn from the students about the material and also how to communicate better with them. The teacher and student in dialogical forms of learning are mutually interdependent (Bailey, 2003).

The programs or commands that Alexa can perform can be understood as 'skills' (Davie and Hilber, 2018), and devices like the Echo Look are part of a suite of skill-building processes with the participant and within the device itself. Amazon overtly marketed the Echo Look as a device that contained knowledge in the form of an expert teacher who could guide the user and, in turn, make it easier to decide what to wear. Machine learning, a key mechanism underpinning the Echo Look, involves training and skill-building through its interactions with user data. Yet, the Echo Look seemed to present information and guidance based on fixed notions that expressed inbuilt biases and ideologies. In the case study of Kenna, she felt that the decisions made by the machine were predicated by a gendered and normative idea of dress. Kenna was excited by many of the features of the device but believed that it could be improved by challenging its own in-built assumptions and the biases of the algorithm, a feature that is common in AI-assisted decision-making processes (Rastogi et al., 2020). Learning from the user might counter the original biases that determined the way the machine analysed data and, if an adaptation was made, it could result in more useful and relevant advice. Participants wanted better dialogue with the machine and hoped the device could learn from them, including being receptive to the elasticity of their preferences in different situations based on the nature of the event and other factors.

In its current form, the Echo Look fell short of a more dialogical approach and was unable to achieve the nuanced assessments and recommendations people were looking for, despite promises for customisation. Participants felt that a dialogical approach that attended to context would not only provide enhanced content but also help to build a form of trust that was not focused on technological implications for security (Chung et al., 2017). This required developing a relationship with the device that would enable the participant to have faith in the recommendations of a device that knew her much in the same way as participants often described, as trusting the opinion of a partner or close friend. The desire for a dialogical approach was then not only about improving advice but also the estimation of both parties in the human–machine interaction as Freire (1970) and Ramis (2018) similarly argued in deconstructing the dual roles of teacher and learner.

The participants who saw potential in the (now discontinued) Echo Look felt that they missed out on an opportunity to learn. Avery, for example, noted Echo Look's preferences made her think more broadly about pairing different pieces than she would have before. She described how she never thought about pairing white trousers with her tops before because she usually wore blue jeans and she continued to explore this style. This process was not limited to focus on feedback on specific outfits. Avery was more interested in what she felt was an overall learning and skill-building process with the Echo Look where she felt it helped her build a set of fashion principles. As she said,

> So it's not just trying to put together an outfit for the sake of scoring higher, but you just might understand the principles that are underlying it . . . for me, it's like being more aware of colour combinations, things that are more complementary towards each other. . . . So having something like Echo

Look helps me figure out what I like together and what I don't, and not necessarily that it just scores better, but it helps me to see what I like better.

This ability to learn together had its challenges. The ways in which participants thought to use the device were largely prompted by the literature that surrounded it and the video marketing tutorials on the Amazon site itself, although one limitation of the pilot study was the relatively short period of time they were able to use the device. Many participants were disappointed that they did not receive verbal feedback from the device and were uncertain about how to initiate discussion, build conversation or establish fluency with the Echo Look. In fact, no one thought to engage the device by asking direct questions beyond the instructions provided in the brochure. When we prompted the participants to ask the device directly the question 'Alexa, how do I look?', the device did reply with the type of information they had hoped that it would provide. But it did not go far enough. The Echo Look did not initiate a discussion or ask them questions. It also only relied upon outfits that were uploaded to learn: other background and contextual information were not included. Participants hoped the device would remember contexts or have sourced information on various events in order to learn their personal preferences and provide more targeted insights. This speaks to a growing body of literature that acknowledges both the rising popularity but also the complicated negotiations of in-home conversational agents in general and Alexa in particular (Sciuto et al., 2018).

Conclusion

In this chapter, we foreground the role of learning in human–machine interactions and the potential applications of dialogical approaches to learning. Despite the aspirations of the device to provide customised recommendations, the Echo Look did not enable or demonstrate a continuous dialogue. Participants enjoyed interacting with the machine and felt that the device had the capacity to provide more individually tailored insights which would make it more indispensable – or at least of greater value – in the process of getting dressed. From the perspective of our participants, the promise of mutual learning where the machine was not only giving recommendations but also itself being trained by humans remains underdeveloped.

The implications of this study are beyond the particular and, in some ways, unsurprising shortcomings of the actual device and its integration into the daily clothing decisions of our participants and other users. Rather, user experiences of the Echo Look have implications for ADM especially for understanding how the meanings and constructions that are embedded in what 'learning' is envisioned and consequently how algorithms are designed and practised through human–machine learning. A more dialogical approach to learning would enable the device to not only learn more about the relevant information but also improve the process of communication. This dialogical relationship then is useful in not only building a better knowledge base of the materiality of the fashion itself but also enhancing its relationship with the people using the device as they navigate the sartorial dilemmas and contexts they traverse in their everyday lives. Our participants articulated a desire for more accurate recommendations and feedback, suggesting a genuine openness towards learning with and from machines in their decision-making. The extent to which machines can listen and learn from nuanced information such as style or context will shape the potential for such automated technologies in decision-making processes into the future.

References

Bailey T (2003) Analogy, Dialectics and Lifelong Learning. *International Journal of Lifelong Education* 22(2): 132–46.

Barrett B (2017) Amazon's 'Echo Look' Could Snoop a Lot More Than Just Your Clothes. *Wired*, 28 April, 18.

Chung H, Iorga M, Voas J and Lee S (2017) Alexa, Can I Trust You? *Computer* 50(9): 100–4.

Cole M (1985) The Zone of Proximal Development: Where Culture and Cognition Create Each Other. In: Wertsch J (ed) *Culture, Communication, and Cognition: Vygotskian Perspectives*. Cambridge: Cambridge University Press, 146–61.

Cole M and Engestrom Y (1989) A Cultural-historical Approach to Distributed Cognition. In: Salomon G (ed) *Distributed Cognitions: Psychological and Educational Considerations*. Cambridge: Cambridge University Press, 1–46.

Davie N and Hilber T (2018) Opportunities and Challenges of Using Amazon Echo in Education. In *14th International Conference Mobile Learning 2018. International Association for Development of the Information Society*. Lisbon, Portugal, 16–18 April 2018. IADIS Press, 205–8.

Freire P (1970) *Pedagogy of the Oppressed*. New York: Seasbury.

Gardner H (2000) *Intelligence Reframed: Multiple Intelligences for the 21st Century*. London: Hachette.

Knight W (2017) Amazon Has Developed an AI Fashion Designer. *MIT Technology Review*. Available at: www.technologyreview.com/2017/08/24/149518/amazon-has-developed-an-ai-fashion-designer/ (accessed 10 June 2021).

Larus J, Hankin C, Carson S, Christen M, Crafa S, Grau O, KirChner C, Knowles B, McGettrick A, Tamburri D and Werthner H (2018) When Computers Decide. European Recommendations on Machine-learned Automated Decision Making. Technical Report. New York, NY: Association for Computing Machinery.

Lave J (1988) *Cognition in Practice: Mind, Mathematics and Culture in Everyday Life*. Cambridge: Cambridge University Press.

Loschek I (2009) *When Clothes Become Fashion: Design and Innovation Systems*. Oxford: Berg.

Luce L (2018) *Artificial Intelligence for Fashion: How AI Is Revolutionizing the Fashion Industry*. New York: Apress.

McGlaun S (2020) Your Amazon Echo Look Is Officially Dead But You Can Score a Sweet Freebie for Your Trouble. *HotHardware*, 30 May [Blog]. Available at: https://hothardware.com/news/echo-look-is-discontinued (accessed 26 June 2021).

Norell N, Nevelson L, Sharaff I, NikolAis A, Courreges A and Tucker P (1967) Is Fashion an Art? *The Metropolitan Museum of Art Bulletin* 26(3): 129–40.

Rahwan I and Simari GR (eds) (2009) *Argumentation in Artificial Intelligence*. Heidelberg: Springer.

Ramadan Z (2019) The Democratization of Intangible Luxury. *Marketing Intelligence & Planning* 37(6): 660–73.

Ramis M (2018) Contributions of Freire's Theory to Dialogic Education. *Social and Education History* 7(3): 277–99.

Rastogi C, Zhang Y, Wei D, Varshney K, DhurAndhar A and Tomsett R (2020) Deciding Fast and Slow: The Role of Cognitive Biases in AI-assisted Decision-making. arXiv:2010.07938 [cs.HC]. https://arxiv.org/abs/2010.07938

Sciuto A, Saini A, Forlizzi J and Hong J (2018) "Hey Alexa, What's Up?" A Mixed-Methods Studies of In-Home Conversational Agent Usage. In *Proceedings of the 2018 Designing Interactive Systems Conference*, 857–68.

Shor I and Freire P (1987) What Is the "Dialogical Method" of Teaching? *Journal of Education* 169(3): 11–31.

Smagorinsky P (1995) The Social Construction of Data: Methodological Problems of Investigating Learning in the Zone of Proximal Development. *Review of Educational Research* 65(3): 191–212.

Strengers Y and Kennedy J (2020) *The Smart Wife: Why Siri, Alexa, and Other Smart Home Devices Need a Feminist Reboot.* Cambridge: MIT Press.

Sun L and Zhao L (2018) Technology Disruptions: Exploring the Changing Roles of Designers, Makers, and Users in the Fashion Industry. *International Journal of Fashion Design, Technology and Education* 11(3): 362–74.

Surur (2020) Amazon Echo Look Discontinued: Turns out Having a Camera Watch You Dress Was Too Creepy Afterall. *MSPoweruser Blog*, 29 May [Blog]. Available at: https://mspoweruser.com/amazon-echo-look-discontinued/ (accessed 11 August 2021).

Vincent J (2017) Amazon's Echo Look Is a Minefield of AI and Privacy Concerns. *The Verge* 27.

Vygotsky LS and Cole M (1981) *Mind in Society: The Development of Higher Psychological Processes.* Cambridge, MA: Harvard University Press.

Woodward S (2007) *Why Women Wear What They Wear.* New York: Berg.

Yurieff K (2017) The Future of Getting Dressed: AI, VR and Smart Fabrics. *CNN Tech.* Available at: http://money.cnn.com/2017/11/13/technology/future-of-fashion-tech/index.html (accessed 20 March 2021).

7

COFFEE WITH THE ALGORITHM

Imaginaries, maintenance and care in the everyday life of a news-ranking algorithm

Jakob Svensson

Introduction

It is Tuesday afternoon and the Daily's newsroom is bustling with people. In a shielded corner in front of the editorial team, journalists, web developers and programmers are gathering for a meeting. It is time for Algorithm Coffee, a meeting to discuss and raise concerns regarding an algorithm automating the ranking and mixing of news stories on the frontpage of the Daily (the Algorithm henceforth). The smell of freshly brewed coffee is spreading from the filled cups people are bringing to the meeting. This week, Lars (fictitious names are used throughout this chapter) from the subscription department is joining. Apparently, the number of new subscribers is down compared to last week. He is thus eager to 'lock' news stories on the frontpage, that is, stories that readers would be willing to pay for. Would it be possible to know for what, and when, a reader would be ready to open their wallet? The programmers attending the meeting are sceptical. They would need someone from the data analytics team to join the meeting to answer these questions. But they only have access to data from users who have logged into the site and have consented to their data being collected (due to European data protection legislation). They are most likely already subscribers, one of the programmers remarks. Perhaps it would be possible to automatically lock a news story, but after it already generated some traffic, a web developer suggests, which would mean only the most popular news stories would be behind a paywall. The social media editor raises some concerns. If she put a blurb with a link to the story on the Daily's Facebook account, and then all of the sudden the story was locked for subscribers only, it would not help to attract new readers from Facebook to the Daily. But maybe this would push interested readers to pay for a subscription, Lars comments enthusiastically. One journalist suggests starting with Stefan's chronicles. His pieces always attract a lot of attention due to his polemic style. The decision is made, and

DOI: 10.4324/9781003170884-10

at next week's Algorithm Coffee they will evaluate the results. Coffee cups are emptied, and people return to their stations in the vast newsroom.

This chapter addresses automated decision-making (ADM) and the role of humans in it by attending to the everyday life of a news-ranking algorithm. The study is set in a Scandinavian news organisation and attends to Algorithm Coffee and other socio-institutional practices emerging through the introduction, development and maintenance of the Algorithm. While algorithms have most often been discussed at the output stage (Klinger and Svensson, 2018), this chapter directs attention to the *input* stage and what goes on *behind* the screen (see also Mansell, 2012). Following Neyland (2019), the aim is to make sense of the everyday entanglements of the Algorithm. Disenchanted with what he labels the *algorithmic drama* – popular concerns about the supposed power and opacity of algorithms – Neyland argues that 'the everyday, humdrum banalities of life are somewhat sidelined' (Neyland, 2019: 3). This is also the case in academic literature, which, he argues, has consequences for our understanding of algorithms. This chapter, therefore, takes as its starting point the questions of how the Algorithm participates in the everyday, composes the everyday and also becomes the everyday in the Daily's newsroom.

The terms 'automation' and 'algorithm' are sometimes difficult to separate. Dourish (2016), for example, highlights how the term 'algorithm' is used metonymically to address the regime of digital automation more broadly. Within the academic literature automation is sometimes approached as a complete tech take-over, something that cannot be avoided, a fait accompli. Andrejevic, for example, argues that we live in an era of a *cascading logic of automation* (2020: 9), referring to how automated data collection leads to automated data processing, which leads to an automated response. This *bias of automation* (Andrejevic, 2020: 21), he argues, pre-empts human decision-making by operationalising large datasets and their collection, through which it is believed that everything is captured from every possible angle. According to Andrejevic (2020: 30), there is a post-social bias as automation attempts to displace social processes with machinic ones, replacing humans as well as human judgement and decision-making.

In contrast to Andrejevic's post-social account of automation, other scholars have proposed that algorithms are *both* social and material processes (Bucher, 2017; Neyland, 2019). By 1986, Noble had already argued that automation – more than merely a technological advance – is a social process. Therefore, algorithms should be studied in their socio-institutional as well as sociotechnical situations and should be approached as unstable and as enacted through people who engage with them, rather than as constrained and procedural formulas (Seaver, 2017; Neyland, 2019).

Imaginaries are helpful for studying socio-institutional practices connected to algorithmic automation. Technology is shaped by imaginaries, values, social and cultural perceptions guiding the development of communication systems (Mansell, 2012: 31). Following Taylor, Mansell defines imaginaries as deeper normative notions and images invoked by how people make sense of practices and how this impacts (in Mansell's case) the Internet. Imaginaries thus influence not only how technology and digital platforms are used, and how they permeate and mediate people's lives (Mansell, 2012: 33), but also how they shape algorithms themselves. Bucher (2017: 40–1), for example, argues that *algorithmic imaginaries* are productive; in that, they influence behaviours around algorithms as well as how algorithms are developed. She shows how such imaginaries are important for, in her case, the moulding of the Facebook algorithm itself. An algorithm is thus a result of human (and not only software programmers) thinking, reasoning and imaginaries. As such, imaginaries – as well as algorithms – are never settled and always up for negotiation (see also Mansell, 2012: 5).

The chapter will, therefore, focus on how algorithmic automation is imagined in the Daily's newsroom and the everyday socio-institutional practices its imaginaries are embedded in. When news workers imagine algorithmic automation at the Daily, they interact not only with the Algorithm but also with each other in their professional capacities. As this chapter shows, these interactions shape the very calculations of the Algorithm itself.

The empirical data for this chapter consist of interviews and observations. The interviews took place during 2018 and 2021 (most at the Daily and the rest at cafés or lunch restaurants). Seven days, over two visits, were also spent at the Daily, both to conduct interviews and sit in meetings and to observe the different actors and sections in the newsroom. Concerning ethics, participants were always aware that I was undertaking research and of my identity as a researcher. Due to a request from the Daily, both individual participants and the Daily itself have been anonymised. A confidentiality agreement with the Daily, among other things, prohibits me from directly revealing the name of the newspaper or the Media Group. Following this agreement, I do not disclose how many people were employed at the Daily, or any details about the organisational structure. However, this is a minor limitation, given the focus of this chapter is human–centred aspects of algorithmic automation rather than to present a case study of a news organisation.

Imagining automation at the Daily

The Algorithm was introduced at the Daily in 2015, with the promise of saving money by automating 'boring', repetitive and mechanical labour such as dragging news stories up and down the frontpage as new news stories emerge. Since then, the Algorithm has been continuously maintained, developed and tweaked to meet certain targets, such as increased subscription rates (as in the introductory scene) or more clicks into the frontpage (i.e. advertisement goals).

At the Daily, the Algorithm's placing and mixing of news stories on the front-page are continuously discussed and evaluated by different actors and according to different logics. These include whether the news mix is representative of the brand, lives up to journalism's democratic mission of providing for an informed citizenry, whether it satisfies commercial imperatives with top-ranked news articles being read (i.e. clicked on), or if it converts readers to digital subscriptions. In other words, in the Daily's newsroom, different actors met, discussed, imagined, experienced and made sense of the Algorithm. The human actors involved are both tech actors and traditional news actors. Most traditional news actors (journalists and editors) are located within the Daily's offices. When it comes to tech actors (programmers, UX designers, data analysts, tech and web developers), most of them are also located at the Daily, but some are based centrally at the Media Group. But actors are not necessarily human (Latour, 1996). The Media Group – most often acting out of the advertisement and the subscription departments – played an important role with its demand for profit. The brand, the Daily with its characteristics and its claimed niche in the nation's news ecology, made its mark in the introduction and development of the Algorithm. And the Algorithm itself can also be considered as an actor.

Algorithms are often understood as problem-solving technologies (Striphas, 2015). They are developed around a problem, on the basis of a corresponding belief that algorithmic automation can solve the problem. For the Daily, this problem was profitability. The number of visits to the website was down, advertising was down and there were even rumours the Media Group wanted to sell the Daily to its competitor. But instead, the Media Group made the newspaper a test bed for algorithmic automation. A team was put together by the web development team at the Daily to launch a new webpage, at a request from the Media Group. This was a rather loose group of programmers, developers and journalists, who came and went depending on both their personal lives (such as paternity leave) and the particular aspects of the frontpage being discussed.

To save money, the Daily needed to do more with less people and to use its journalist resources better. As one journalist puts it, to not have them 'manually drag news stories up and down their online front-page'. Thus, the idea that these tasks could be automated by an algorithm was born. The imaginary that algorithmic automation is 'labour-saving, making it possible to do more things without employing more people' and that 'journalists should do journalism stuff and not drag things up and down the front-page' was prominent in the interviews.

The task of automatically sorting and mixing news on the frontpage was not directly connected to how journalists at the Daily imagined their democratic purpose. The Algorithm was developed so that it did not interfere with journalism's mission to give all readers/citizens the same balanced news mix. At least this is how it was made sense of, as exemplified in an interview with one of the web developers:

> We have an algorithm that controls which news reaches the readers. We have chosen not to personalise this too much. It is an important aspect of a democracy that no matter who you are, you should get the same news mix.

Hence, as long as it did not interfere too much with the imaginary of journalists providing readers with news to fulfil their civic function in a representative democracy, the Algorithm could be justified. At the same time, it became apparent that journalism's democratic mission was of secondary concern for the Media Group, where profit-making took centre stage in the imaginaries surrounding the introduction and development of the Algorithm. The newsroom setting is nonetheless important for understanding the imagined boundaries of automation and how these were negotiated. I return to this later in the chapter when discussing the Algorithm as *editor-led*.

When it comes to profit-making, so-called *native* advertising (integrated into the news flow like news stories) was gaining ground in relation to *display* (based on page views) which is still the major income in terms of digital advertisements. The Daily had gone from being 95 to 80% advertisement financed, with digital subscriptions going up. Material for paying subscribers was thus becoming increasingly important, as was locking news stories to convert readers into paying subscribers. During the time of study, the Daily started to experiment with adding a parameter to the Algorithm that would automatically lock news stories for subscribers only, mainly based on how much traffic the story had generated (as exemplified in the introductory scene). It is interesting that this parameter of the Algorithm went under the nickname *the Oracle*. This invokes the connection between imagination of problem-solving through algorithmic automation to wizardry and fortune-telling, and how oracles and shamans had a sensibility towards the future. Pasquale (2015) connects the allure of algorithms to the ancient aspiration to predict the future but through data instead of a crystal ball or intoxicating fumes as for *Pythia*, the famous oracle in Delphi.

The Oracle parameter suggests that it is possible to predict readers' intentions and to anticipate their behaviour based on the data traces they leave behind when logged in to the Daily webpage. This resembles what Morozov (2013) critically labels as *technological solutionism*, recasting complex situations as neatly defined tech problems with computable solutions. Through their skills in computing and access to data, programmers are thought to help solve all kinds of problems in magical ways (see Svensson, 2021). Nagy et al. (2020), therefore, refer to technology as *a magical panacea* for all problems in almost all walks of life. Tech is the trick that makes a problem disappear or become solved. Comor and Compton (2015) refer to the belief in technology as inherently powerful as a *technological fetish*. Indeed, interviews at the Daily suggest a kind of *jump-on-the-bandwagon* mentality that the newspaper had to 'have these systems', 'build on these techniques' and 'think in these ways'.

At the Daily, interestingly the imaginary of technological solutionism was mostly championed by the journalists. Some in-house programmers even complained about journalists expressing wishes for technological solutions that were completely unrealistic, as in the following example (emphasis in original):

> If someone who is *not* technically knowledgeable should think about a solution that . . . "there should be a button there and when you press the button it should come up with exactly the stuff you like" (mimicking a journalist, authors remark) . . . and then you are like . . . from where does the stuff you like come from? And he answers, "from pressing the button" . . . but if there should be a button that selects articles you like, then there must be a model behind it, some really advanced AI.

Such tech solutionist imaginaries created stress for the in-house programmers. At one meeting, held at 3 o'clock in the afternoon, one editor demanded that a new linking function (automatically providing relevant links to a published news story) should be up and running on the frontpage two hours later (at 5 o'clock). One of the programmers complained that this would not be possible and would need a novel tagging hierarchy of all news stories in the database. The stressed editor responded, 'how difficult can it be, you know these things, just do it!' At the meeting, the editor also underlined how the Algorithm would do a better job at providing these links as it would not be tainted by human editors' biases, favouring (as well as having more knowledge about) their own news stories. Indeed, algorithms are often accompanied by imaginaries of being more neutral and less biased than human actors (Klinger and Svensson, 2018).

For example, during the Algorithm's introduction in 2015, programmers worked with an algorithmically automated frontpage alongside an editor who manually rendered a frontpage and then compared the two. By showing similarities between the two pages, programmers could eventually convince editors of the benefits of algorithmic automation. Interestingly, they also used the Daily's brand, which holds a particular niche in the Scandinavian news ecology, in this argument. The comparison allowed programmers to show that the gut-feelings of different editors varied. Therefore, the brand influenced the adoption of the Algorithm, since algorithms are accompanied by imaginaries of objectivity, and as being detached from human bias. In the words of a web developer:

> Before you could notice who had had the editing shift. If there was a lot of international news on the front-page, then we knew who the editor had been. Now it is a more coherent and recognizable product.

The Algorithm was nevertheless labelled *editor-led*. As one programmer explained to me: 'We are not automating it completely, because there is an editorial decision behind the news value, journalists have the final say'. Programmers at the Daily were sympathetic towards journalism and its role in a democratic society. At the meetings, there were constant reminders that the Daily should be 'journalist-controlled', and that the in-house programmers should get their instructions from the editors and should support their requests. In fact, most of the in-house programmers had chosen their workplace because they valued journalism, as one commented:

> I am quite interested in societal matters you know, and it is fun to work with a product you care about, a luxury to think that what you do is good . . . and to feel the pulse from the newsroom and journalism.

It is perhaps not surprising that programmers who chose to work in a news organisation appreciate journalism. It was probably these circumstances that made the Algorithm viable for this particular newsroom context in the first place. The way the Algorithm was developed was therefore partly a result of prevalent journalistic values. The journalists at the Daily were back-patted, and their unique function in democracy was often emphasised. But sometimes the journalists needed to be disciplined, often with reference to how a supposedly more neutral algorithm could do a better job in ensuring a coherent news mix/brand.

So what decisions were automated by the Algorithm, given that it was co-produced by imaginaries connected to both journalism and automation? Many algorithmic calculations were executed to produce the automated rendering of the frontpage. News stories were selected from a pool to create the news mix that was so important to the brand. Then individual news stories were valued in relation to the existing mix of stories on the frontpage in order to place them properly. There are rules that determine that there should be a certain number of news stories from different news categories/genres, to ensure readers would recognise the Daily/the brand. The different parameters of the Algorithms calculations revolved around time (referring to the latest news and the longevity of a news item), news value (in terms of 1 to 5), subscription conversion (if a news item converted many readers to paying subscribers, it was pushed higher) and popularity (in terms of clicks/page views). The size of the picture, headline, preamble and what links should accompany the story are rendered automatically (based on how the news item is tagged and valued). There are also tagging hierarchies so that news items dealing with the same story (e.g. Spanish wildfires) could be grouped together (maximum three news items in one story) and then even larger *super-stories* with many stories underneath it (such as the Taliban take-over in Afghanistan). Interestingly, time and news parameters were manually reported by an editor, while the other parameters were automated. Hence, referring to the Algorithm as *automating* news-ranking and mixing is only partially true. Not all parameters of the algorithmic computation were automated. This nuances Andrejevic's (2020) arguments that we live in an era of a cascading logic of automation and that automation comes with a post-social bias. When it comes to ADM, the study shows how intricate and entangled this is with the everyday socio-institutional practices in the newsroom.

This example emphasises that algorithms are co-constituted (Seaver, 2017) and not solely an outside force. Imaginaries are productive and they cannot be disregarded as fetish or false beliefs (Bucher, 2017: 31). At the Daily, the journalists' democratic mission, and their imaginary of giving citizens what they need, were coded into the Algorithm through the news value parameter. Journalism's rule of breaking news was coded in through the time parameters. The Media group's quest of profit-making was coded in through the parameters of subscription conversion and popularity. In other words, imaginaries of journalism, its role in a democracy, as well as automation (tech solutionism) were productive in the introduction, development and maintenance of the Algorithm.

Socio-institutional practices in the everyday life of the Algorithm

The Algorithm is not static. Since its introduction, its parameters have been continuously supervised and fine-tuned (i.e. maintained), informed by imaginaries, tensions and negotiations behind the frontpage of the Daily, highlighting how an algorithm is most often, if not always, in progress. Of particular interest here are meetings and situations where actors within the Daily gathered to collectively make sense of, maintain, tweak and thus co-develop the Algorithm. What happens when people enter the stage of the 'drama' of algorithmic automation? At the Daily, it became apparent that news workers also interacted with each other through the Algorithm. The Algorithm became a new battleground but for old conflicts in the newsroom. And it was the in-house programmers who ended up mediating the – often contradictory – imaginaries of journalists and subscription and ad sellers. This influenced the Algorithm's calculations.

There were times when journalists questioned the Algorithm, for example why a certain news story was at the top of the site and why a certain news story got a bigger picture than another. After a while, they learned how the Algorithm functioned. For example, they realised that if they wanted their story to get a bigger picture, they had to value it at least as a 3 and with a lasting news value. Hence, editors started to manipulate the Algorithm, or 'tame' or 'massage' it, as the head of digital development expressed it in an interview:

> My job is to not *tame* the Algorithm, but to make us work in a way that the Algorithm does what we believe is the best. . . . We do not manipulate the Algorithm, but we can massage the Algorithm [my emphasis].

In addition to getting feedback on the Algorithm, one of the purposes of Algorithm Coffee was to 'discipline' the journalists trying to 'massage' it. This disciplining also revolved around tensions in the newsroom connected to the brand and its supposedly unique news mix. For example, a sports news story (fictitious example) could not be given more than a 3 in news value in order to ensure a perfect news mix for the brand. Tensions around this value – which sports journalists were obviously not happy about – were often played out in front of the in-house programmers in their corner in the newsroom. Different actors came to the programmers and complained about the Algorithm as they championed their interests, for example in ranking sports news higher on the front page.

Another tension revolved around how much space ads should be allowed on the frontpage. Since 'there was no one in charge of balance between the advertising, market and the editorial staff', as one programmer told me, the programmers 'had to mediate between these groups'. Tensions between journalists, ad and subscription departments are not new to newsrooms (see Asp, 2014). But they are recast and revived in this situation of algorithmic automation. Not only were programmers perceived as tech solutionist problem-solvers, but they also functioned as mediators between different groups as the Algorithm became a new battleground for old conflicts in the newsroom.

All this had a bearing on how the Algorithm was developed and maintained. During the introduction of the Algorithm, the most apparent tension was that between programmers and journalists. This tension concerned automating something 'that had been someone's baby, to do the front-page', as one programmer phrased it. At the Daily, programmers talked about journalists 'not being used to technology having a big impact on their everyday work'. Journalists themselves mentioned their fears 'having built a career and pride manually controlling the front-page' they were now confronted with a frontpage run by 'IT-boys without editorial experience'. One journalist described this as a feeling of their work and themself being 'worthless when you can just be replaced by technology'. Andrejevic (2020: 10) explains such anxieties in terms of automation as cascading, whereby automation will supplant not only human labour but also human autonomy.

But automation was not really cascading at the Daily. The anxieties and workplace relationships that the Algorithm revived were addressed over a cup of coffee. The idea of Algorithm Coffee was first born to channel and negotiate tensions between different groups and to discipline journalists trying to massage the Algorithm. Programmers, developers, journalists, advertisers and data analysts needed to meet and talk to each other when maintaining and tweaking the Algorithm and its calculations. The coffee was a bribe to get people to come to discuss 'the week with the Algorithm', as one of the programmers who had initiated it explained to me. Significantly, the programmers at the Daily set this up to mediate the tensions created by the different imaginaries about what algorithmic automation could and should do. On the occasions that I attended Algorithm Coffee, there were questions raised concerning how the Algorithm performed to different targets, the motives for the placement of news stories, and whether a parameter should be fine-tuned or not, on the basis of what data, or what data could be produced if the Algorithm was tweaked in a certain way.

During a meeting just after the elections to the Swedish Parliament (which had resulted in no clear majority), tensions between the newsroom, ad and the subscription departments were foregrounded. Johan from the advertisement department was excited because they had a company interested in buying the whole frontpage for a lot of money: 'If we have this ad that you need to click through, before entering into the frontpage, we would be able to make X amount'. Lars from the subscription department was not convinced. He questioned the long-term business strategy of this: 'Readers do not come to the Daily to see a huge advertisement, they come to read quality news and update themselves on what is going on in the country and the world'. Lars was seconded by one of the editors who added that this was not an optimal day for that kind of experimentation as the election of the Speaker to the Parliament was held later that afternoon: 'We know readers will surf into the site to get the latest on that'. But he also turned directly to the participating programmer to highlight that this story should not be locked for subscribers only, even if it would generate a lot of traffic into the site. 'This is important news for citizens in our country; everyone should be able to access the latest on this'. Johan was still not convinced: 'we need to make a profit here, even after an election, and right now it is the ads that bring in the revenue'.

Algorithm Coffee seems to have worked in relation to the initial tensions between programmers and journalists. The Daily now has a 'digital first thinking', as one journalist phrased it. 'I love the Algorithm' exclaimed another editor adding that 'we get more time to be editors now'. In-house programmers say they feel journalists trust them more and one journalist told me she is not 'afraid any longer'. The Algorithm had thus become *a matter of care*. Puig de la Bellacasa (2017) argues for the significance of care when thinking and living in more-than-human worlds. Following Tronto and Fischer, she defines care as things we do to maintain, continue and repair our world (Puig de la Bellacasa, 2017: 3). That news workers express love for the Algorithm, and meet regularly to discuss and maintain it, is thus an example of care for the Algorithm and how it worked. Algorithm Coffee thus functioned as a form of *care work*, which apart from its affective side also entailed concrete work of maintenance (Puig de la Bellacasa, 2017: 5).

It was not only the Algorithm that was cared for during Algorithm Coffee, but also workplace relationships and professional identities were maintained, continued and repaired during these meetings. As this chapter has shown, the Algorithm challenged workplace relationships at the Daily and awoke old conflicts in the newsroom. It, therefore, had to be moulded or cared for in order to avoid creating too much disruption. Care here is thus not only for the technology and how it functioned but also for all those concerned with the technology (see Puig de la Bellacasa, 2017: 48). In this sense, Algorithm Coffee underlines a double significance of care, both as everyday labour of maintenance, and taking care of a thing, and as such to remain responsible for its becoming (see Puig de la Bellacasa, 2017: 43). Algorithm Coffee is thus about making time for care in more-than-human worlds. As Puig de la Bellacasa outlines, 'we cannot afford to obscure the actual more laborious and situated conditions in which care takes place and by which its agencies circulate in interdependent more than human relational webs' (2017: 24).

At the Daily, journalists have accepted the Algorithm, with its imaginary of freeing resources for them to do 'more important work'. In the interviews, allowing for the Algorithm to be editor-led, and thus only semi-automated, with journalist and tech actors working tightly together to tweak and maintain the Algorithm, often over coffee, were emphasised as factors behind the editors' and journalists' acceptance and care for the Algorithm.

Conclusion

When approaching algorithms as organic, incomplete and situated (Bucher, 2017; Seaver, 2017), socio-institutional practices of maintenance and care take centre stage. Through directing attention to how algorithmic automation is imagined by actors in the Daily's newsroom, it is possible to empirically approach the Algorithm as an unstable object (Seaver, 2017). As this chapter has shown, algorithms are enacted through people that engage with them and by practices which do not heed a strong distinction between the technical and non-technical, but rather blend them together (Seaver, 2017). The chapter has highlighted how the Algorithm is most often in progress, in constant development as the weight of the Algorithm's different parameters are continuously maintained and fine-tuned and new parameters are added.

This maintenance and fine-tuning were often informed by conflicts and relationships between different groups at the Daily. These had a direct bearing on how the Algorithm developed. Accounting for its newsroom setting is pivotal to understanding the Algorithm – its introduction, development and maintenance.

Imaginaries of algorithmic automation revolved around tech solutionism and how it could solve the Daily's profitability problem by automating tasks previously done manually by an editor. It was claimed that this was not directly connected to journalisms' higher purpose, and hence, such automation could be justified because editors would remain in control of time and news value parameters. Indeed, the Algorithm was labelled editor-led, despite having been sold as time and labour-saving in its capacity to *automate* editorial decision-making. The chapter has thus shown how different actors in the newsroom co-developed, negotiated and made sense of algorithmic automation, not the least through allowing human editors to oversee some of its parameters. We are thus not talking about complete automation. This raises questions about the term automation as such. What is meant with automation when in this case we refer to a hybrid between some parameters being automated, while others manually entered, and all parameters being open for continuous negotiations, constant tweaking and maintenance by human actors having coffee together on a regular basis? Indeed, algorithmic automation is not a fait accompli, first it needs to be imagined, then negotiated, maintained and cared for.

Interestingly, the Algorithm also functioned as a medium through which human relationships at the Daily were cared for. Old tensions and workplace relationships were recast and revived through the Algorithm. Algorithm Coffee is a practice of caring for workplace relationships in a situation where the programmers ended up mediating tensions between contradictory imaginaries around algorithmic automation. Indeed, human intervention, relationships and socio-institutional practices are all over the Algorithm and its calculations. This shows that human and machine decision-making cannot be separated and need to be studied in tandem. Furthermore, by attending to everyday socio-institutional practices surrounding automation, automation becomes less opaque and less mysterious. The Algorithm and its five different parameters were not rocket science, not even for me as a social scientist.

The contribution of this chapter is thus to turn the question of automation around. Not only should we ask how ADM impacts everyday life, but also how everyday life – of both humans and non-humans – impacts automation. Everyday practices of maintenance and care, such as having coffee to discuss algorithmic automation, are indeed important. Such practices are also important for imagining automation and making such imaginaries matter (in the dual meaning of the word), in this case through maintaining and tweaking the parameters of algorithmic automation as well as adding new ones.

References

Andrejevic M (2020) *Automated Media.* New York: Routledge.

Asp K (2014) News Media Logic in a New Institutional Perspective. *Journalism Studies* 15(3): 256–70.

Bucher T (2017) The Algorithmic Imaginary: Exploring the Ordinary Affects of Facebook Algorithms. *Information, Communication & Society* 20(1): 30–44.

Comor E and Compton JR (2015) Journalistic Labour and Technological Fetishism. *The Political Economy of Communication* 3(2): 74–87.

Dourish P (2016) Algorithms and Their Others: Algorithmic Culture in Context. *Big Data & Society*, July–December: 1–11.

Klinger U and Svensson J (2018) The End of Media Logics? On Algorithms and Agency. *New Media & Society* 20(12): 4653–70.

Latour B (1996) On Actor-network Theory: A Few Clarifications. *Soziale Welt* 369–81. Available at: www.jstor.org/stable/40878163 (accessed 14 April 2020).

Mansell R (2012) *Imagining the Internet. Communication, Innovation and Governance.* Oxford: Oxford University Press. https://doi.org/10.1177/1464884919861598

Morozov E (2013) *To Save Everything, Click Here. The Folly of Technological Solutionism.* New York: PublicAffairs.

Nagy P, Eschrich J and Finn E (2020) Time Hacking: How Technologies Mediate Time. *Information, Communication & Society*, May 2020. https://doi.org/10.1080/13691 18X.2020.1758743

Neyland D (2019) *The Everyday Life of an Algorithm.* London: Palgrave Macmillan.

Noble DF (1986) *Forces of Production: A Social History of Industrial Automation.* New York: Oxford University Press.

Pasquale F (2015) *The Black Box Society. The Secret Algorithms That Control Money and Information.* Cambridge: Harvard University Press.

Puig de la Bellacasa M (2017) *Matters of Care. Speculative Ethics in More Than Human Worlds.* Minneapolis: University of Minnesota Press.

Seaver N (2017) Algorithms as Culture: Some Tactics for the Ethnography of Algorithmic Systems. *Big Data & Society*, July–December: 1–12.

Striphas T (2015) Algorithmic Culture. *European Journal of Cultural Studies* 18(4–5): 395–412.

Svensson J (2021) *Wizards of the Web. An Outsider's Journey Into Tech Culture, Programming and Mathemagics.* Göteborg: Nordicom.

8

EVERYDAY AI AT WORK

Self-tracking and automated communication for smart work

Stine Lomborg

Modern organisation and employee monitoring are inextricably connected, writes Kirstie Ball in her 2010 review article on workplace surveillance (Ball, 2010). Ball notes that employee monitoring was originally legitimised by way of the contract between the organisation and the employee concerning the exchange of labour for payment. From the productivity and efficiency logics of 20th-century Taylorism to the development and raison d'être of the human resources (HR) discipline, investing in employees' wellbeing and development to retain them and keep them committed to advancing company goals, performance measurement and employee monitoring are key to optimal organisational functioning. In one sense, an organisation has a legitimate and contractually sanctioned interest in ensuring that employees perform the job they are paid to do and thus in leveraging technologies for oversight and control of workplace productivity, etc. But the legitimacy of such interests is challenged when monitoring of the working employee is not limited to the workplace but is 'always-on' and pervades other domains of daily life (Hull and Pasquale, 2019; Moore et al., 2017).

This is often the case with digital tracking enhanced with artificial intelligence (AI). With work organised and performed with or through digital media, ubiquitous employee tracking increasingly becomes the reality; it sits uneasily in tensions with privacy, control and personal autonomy (Calvard, 2019). Recently cases of and debates about 'gig work' have forcefully reflected such tensions. Indeed, we may think of these as extreme cases of digital and datafied work (Delfanti and Frey, 2021; Rosenblat and Stark, 2016). But the same tensions can be found in initiatives of white-collar 'smart work', where AI-enhanced technologies for tracking and analysing an individual's work habits are implemented not only to enhance productivity but also to enable better work/life balance for a stressed-out workforce.

This chapter considers office-based workplace tracking, enhanced with machine learning, as a contemporary form of everyday AI, offering an analysis of forms of automation that go unnoticed and integrate seamlessly in daily life. Emerging

DOI: 10.4324/9781003170884-11

automated decision-making (ADM) systems perform myriad functions in society. Some of them have the capacity to disrupt businesses and change living circumstances, but others are mundane and perceived as relatively unproblematic, even beneficial for citizens (Pink et al., 2017). Automated and AI-enhanced systems have largely gained attention in social science research and public debate in relation to cases of failure, where specific ADM systems typically developed for or implemented in the public sector are shown to exacerbate existing inequalities and injustice or even violate basic human rights (Eubanks, 2018; Crawford, 2021). The exposure of some of these fundamental problems has led to the retraction of some of these systems.

We seem to speak less about automated decision-support systems that are 'mundane'. Here, I mean mundane not only in the sense of being commonly implemented, encountered and engaged with in everyday life but also perhaps not really noticed as we go about our daily business. Such systems are sometimes referred to as 'everyday AI' in the tech industry (Norvig, 2012) and their uses and implications are crucial to study, because these systems tell us important stories about how algorithmic decision-support, derived from applying machine learning to vast amounts of data, currently becomes usefully embedded in organisations and society at large, and how it might become meaningful for individual users of these systems.

I treat 'everyday AI' as a research perspective on various applications of AI that are built into systems we encounter on an everyday basis. This perspective prompts us to address how AI-powered insights and algorithmic decision-support systems are made meaningful as they assist users in getting things done, finding relevant information, and so forth on a daily basis. Everyday AI includes, for example, recommender systems, customer service chatbots, search engine algorithms, smart assistants, digital self-tracking services, etc. I specifically explore productivity and wellbeing self-tracking systems that are being adopted at workplaces in many parts of the world today, understanding such systems in terms of what Deborah Lupton (2016) has called 'pushed' and 'imposed' kinds of digital self-tracking of employees. These self-tracking systems, delivering analytics-based insights on individual workers' productivity or wellbeing, offer emblematic examples of the way digital media intensify the quantification, datafication and automation of work at the level of daily practices (Moore, 2018; Sánchez-Monedero and Dencik, 2019). They are also key examples of how tracking and everyday AI at work might erode the boundaries we typically draw between the personal and the professional domain (Gregg, 2018).

I explore self-tracking at work as a communicative practice to discuss how digital decision-guidance systems (Yeung, 2017) meant to enhance work productivity and wellbeing are being embedded in and contribute to shaping practices of work. In doing so, I draw on examples from qualitative empirical work conducted over the past two years on the management of work and personal life from the perspective of self-tracking of health and productivity. Taking a communicative approach broadens our understanding of what is actually at stake in self-tracking. As pragmatist communication scholar James Carey (1992 [1989]) reminds us, communication is both information and ritual. Unpacking the ritual dimensions of communication, Carey argues that on top of transmitting messages, every act of communication speaks to the social order, participation structures and relevant identities of the context in question. Following from this, communicative analyses put the spotlight on the role of interpretation, ordering and meaning in the specific context of self-tracking, and how tracking enables specific forms of participation relative to context.

My empirical work is situated in Denmark, one of the most digitised countries in the world and a welfare nation context characterised by relatively high trust in organisations and institutions, a historically strong focus on worker rights and a comparably privileged workforce. Here, my explorations will centre on 'smart work' as epitomised by the AI-enhanced system for self-tracking among so-called 'white-collar' workers, Microsoft MyAnalytics. My empirical material is collected as part of a broader study of how digital communication contributes to blurring the boundaries between the personal and the professional domains. The overall study has a broad outlook on self-tracking at the workplace in relation to digital work habits, health and wellbeing and work/life balance. I have collected product presentation videos, press briefs and reviews and interviewed 16 office workers and managers from a variety of public and private organisations in Denmark about their experiences of and reflections about tracking themselves with MyAnalytics and other digital tools, as well as the possible adjustments of core (work) habits that the insights gained with such self-tracking tools have enabled. Some of the interviews were conducted in late 2020 and early 2021, a period when Denmark was under a second lockdown owing to COVID-19. During the periods of lockdown, a large part of the white-collar workforce was working digitally and mainly from home. Needless to say, the ongoing COVID-19 pandemic has pushed discussions of the consequences of tracking and digital work to the forefront in shaping good working futures and working lives.

In the following, I will discuss the continuous, mutual shaping of everyday AI and usage practices through communicative exchanges in these systems. I apply a communication perspective to foreground an analytical sensibility to the diverse, contextually grounded empirical uses of automated decision-support systems that speak to current debates around digital paternalism, micro nudging and concerns over AI and constraint of human agency.

Self-tracking as a case of everyday AI at work

Self-tracking applications typically rely on analytics enhanced by machine learning used on the data input to the system. In the context of work, this is for instance the case with certain forms of self-tracking for healthy workers packaged in corporate wellness programs, such as Fitbit Care, and in productivity and smart work applications that use AI to mine employees' digital traces to inform, target and optimise their work habits, such as Microsoft MyAnalytics, which is the focus case of this chapter. Such self-tracking systems, while certainly contested, are largely implemented in workplaces as part of HR initiatives to support workers' healthy lifestyle, individual work pressure and stress and work/life balance. And they are often *pushed* or *imposed* on users (Lupton, 2016) who are strongly encouraged or provided with clear incentives to take up the offers of self-tracking at work to make themselves better, fitter and happier employees.

As, for instance, Phoebe Moore (2018) and Kate Crawford (2021) have shown, self-tracking at work may be seen as a contemporary addition to historical practices of tracking employees. Henry Ford's streamlining of the factory assembly line and Frederick W. Taylor's scientific management in the late 19th century were geared to control and optimise worker movements for maximum efficiency. Electronic performance management and annual appraisal of individual workers' productivity gained traction in the mid-20th century. The past 30 years or so have marked an increased focus on individual motivation and responsibilisation through corporate identity and wellbeing initiatives, now enhanced with digital trace data in so-called 'people analytics' (Hull and Pasquale, 2019).

Self-tracking applications at work are a continuation of such practices, yet present themselves as automated decision-guidance tools (Yeung, 2017) strongly geared towards the self-optimisation of the individual employee. A prominent example of such technologies targeting workplace wellness and productivity, Microsoft MyAnalytics, uses the employee's data from Office365 to help the individual employee to work smarter and engage in so-called 'deep work'. The idea of deep work was popularised in a bestseller by academic Cal Newport (2016) in his self-help guide to getting more work done without overworking, sustaining a sense of work/life balance while also being immensely productive and successful at work. MyAnalytics, along with other AI-enhanced technologies for self-tracking at work, is marketed and implemented as a means for individual workers to track themselves and enable 'the good work life'. Yet such technologies also constitute new forms of bossware that enables organisational management to track employees at greater scale, in more depth, and across contexts.

Previous studies of self-tracking at work have focused, on the one hand, on what works and what doesn't in terms of using self-tracking to enhance efficiency and wellbeing at work (Elmholdt et al., 2021; Winikoff et al., 2021) and, on the other hand, on theoretical and critical assessments of what might be the implications of digital tracking of employees in terms of worker rights, data-based governance and loss of autonomy (Moore, 2018; Hull and Pasquale, 2019; Ajunwa et al., 2016; Till, 2019). I advance such efforts, trying to bridge perspectives by bringing analytical nuance and contextual sensitivity to how self-tracking at work *communicates* about work and workers by datafying and scoring specific types of work, and how individual employees communicate back to the system by ignoring it, or by appropriating it in ways that are meaningful and perceived as useful from the vantage point of their everyday work lives.

In setting the scene, I take inspiration from Guyard and Kaun (2018), who write specifically about digital distractions at work. In their empirically informed analysis, Guyard and Kaun (2018) studied the implementation of 'workfulness', a managerial strategy for governing collective behaviour at a telecom company in Stockholm in response to 'digital distractions in the workplace'. They suggest this managerial regime of workfulness is guided not by invoking expectations of rational self-control but by top–down stimulus-control: for instance, by banning emails in certain time slots to control employees' impulses to be always on and help them to work in a more focused manner. While MyAnalytics promotes similar ideas of workfulness, it appeals more to the individual employee's rational self-control, hence its base in self-tracking. MyAnalytics presents a case of using data-driven insights to help individual office workers to work smarter, providing suggestions for how this may be done, while granting the individual not only more flexibility but also more responsibility in terms of acting on the provided analytics insights. This responsibilisation can be traced to neoliberal governance structures of optimisation documented in the past decades across domains (see, e.g. Kristensen, 2022; Lupton, 1995).

Microsoft MyAnalytics was introduced in 2016 as an AI-enhanced self-tracking application built on top of the Office365 system. It can be considered fairly mundane and embedded in the daily work grind, thus making a good case of everyday AI. If your workplace uses Microsoft 365, a standard software package for office work, chances are that you are receiving weekly notifications about your working habits, because it has MyAnalytics as a default add-on since 2020. If you have it, maybe you have opened the weekly digest emails it sends and checked out the messages – maybe you have just let them pass in the steady inflow of emails in an already overgrown inbox or deleted them altogether.

MyAnalytics is a system that delivers analytics insights on individual work patterns and behaviour by way of data about the individual's use of the office email client, Microsoft teams, calendar, skype and other data that can be captured within the Microsoft ecosystem. This is done to enable the individual employee to work smarter, whether through enhancing focus time or collaborations with others, and by offering prompts to regulate email behaviour, and by extension, work and leisure time.

In the fall of 2020, Microsoft launched Productivity Score, an extension of the data-driven insights gained in MyAnalytics, to enable organisational management to track the organisational 'adoption of key features' and a declared vision of supporting a future of remote work (from the home). Originally, it involved individual-level data from MyAnalytics, but after a strong public pushback, Microsoft retracted this feature and announced that the Productivity Score would only deal with aggregate level data and not be used for scoring individuals (Spataro, 2020). In early 2021, Microsoft launched a further initiative along similar lines, Microsoft Viva, which is a platform integrated with Microsoft Teams to help businesses with remote work by serving as both intranet, point for connecting with colleagues on collaborations, learning and internal education modules, and tracking individual work habits and moods to deliver personal insights and organisational – aggregate level – insights. In addition, a patent has been filed to use data derived from facial and bodily expressions, time of day, room temperature and so to score the efficiency and quality of meetings and to give wellness recommendations based on data from, among other things, tone of voice in written and audio recorded communications, time spent on emails and keystroke pressure. Hence, the self-tracking enabled in MyAnalytics may be seen as one piece in the development of a larger data-driven system for productivity measurement and optimisation aimed at consolidating Microsoft's market position in the future of work.

Deborah Lupton has described the 'function creep' when self-tracking turns from an individually oriented and voluntary or pushed form into an asset for exploitation by others. She writes:

> the initial incentive for engaging in dataveillance of the self comes from another actor or agency. Self-monitoring may be taken up more or less voluntarily, but in response to external encouragement or advocating rather than as a wholly self-generated and private initiative. In pushed self-tracking, those who are advocating for others to engage in these practices are often interested in viewing or using participants' personal data for their own purposes.
>
> *(Lupton, 2016: 107)*

MyAnalytics is a typical example of such pushed or imposed self-tracking, embedded in the workplace to derive 'AI-powered insight' to foster smart work and help workers improve their working habits on a day-to-day basis. My aim in the following is not to offer a systematic empirical analysis of the experiences and reception of MyAnalytics. Rather it is to use the mundanity of this AI-enhanced system for self-tracking of white-collar employees as an anchor for discussing the possibilities of everyday AI for cultivating a good work life. In doing so, I seek to address the kinds of small-scale, but very significant, challenges that AI-powered decision-support might help people address in everyday life: organising work and managing one's time.

Microsoft MyAnalytics as automated communication

I approach MyAnalytics as a mundane case of pushed and sometimes imposed self-tracking at work from the perspective of communication, building on previous work, where media scholar Kirsten Frandsen and I developed a heuristic of self-tracking as communication with the system, the self and the social world (Lomborg and Frandsen, 2016). Added to this, system-to-system communication denotes the processes by which data from various sources are combined to·produce new insights beyond the meanings produced by users.

The self-tracking enabled in MyAnalytics falls in three consecutive communicative steps. First, Microsoft MyAnalytics, like any kind of everyday AI, is premised on users communicating with digital systems, whereby they leave digital traces to be picked up and made sense of. Most self-tracking applications in the personal domain rely on users' active, voluntary and purposeful data input in their tracking application or wearable device. I have to turn on and wear my Fitbit to enable its step counting or sleep tracking; I actively mark when my period starts and stops in the Clue app. In contrast, MyAnalytics draws its data for self-tracking from a vast data resource that the user has generated for other purposes, namely for getting work done. It encompasses data on attended meetings as indicated by the employee's Outlook calendar, emails written and received in the Outlook mail client according to time of day, calls taken on Skype for Business, collaborative task completion in Teams, etc.

The tracking of the employee with MyAnalytics, therefore, goes largely unnoticed. It is automated, as data are captured automatically from digital systems that are indispensable at work and that the user has not chosen. In addition, MyAnalytics is now turned on by default when using Office365, although it can be actively turned off either by the user or by management decision. Turning it off means the employee will not receive analytics feedback; it does not seem to imply that data capture and analytics processes based on one's data are terminated. Indeed, in MyAnalytics, data are 'taken by', not 'given to' the system (for a discussion of data capture, see, e.g. Kitchin, 2014). MyAnalytics does not track the quality of the work delivered, the informal face-to-face communication outside of formal meetings where new ideas may emerge and challenges solved, or the specific personal circumstances that make particular work arrangements meaningful (e.g. a working parent of small children who checks out early in the late afternoon when care duties typically take centre stage and then logs back on at night to finish the day's work). It is at best what media theorist Mark Andrejevic (2020) has called a 'fantasy of total information capture'. What is captured and thus defined as work is that which is measurable in Office365 (which appears to be increasingly under scrutiny for furthering data-driven analytics and insights as Microsoft prepares its bid for the future of work). By determining what gets to count, MyAnalytics communicates to the user what is important at work, and by extension, what is the expected work ethic and ability to self-regulate in order to be a good employee.

Second, MyAnalytics uses machine learning to analyse the input data according to four main categories of information: time spent on email, time spent in meetings, email activity as mapped onto the time of day and information about communicative connections with important contacts and collaborators in the organisation, along with analyses that correlate these categories, such as the employee's email use during meetings. The analytics performed can be understood as system-to-system communication where information is parsed and cross-referenced automatically to produce meanings about users in ways that are very different from human meaning-making.

In the third step, feedback is delivered by way of automated communication about the key data-driven insights from the system to the individual worker. This feedback takes the form of automatically generated weekly digest emails, 'Your week in review', which shows basic statistics in simple visualisations and automated suggestions of how to tweak or optimise one's digital work behaviour to 'get more focus time' (to be automatically booked in the individual work calendar) or mentally or physically 'recharge the batteries' (e.g. by avoiding email after normal work hours and disconnecting from work). These insights can be explored in more detail on an individual dashboard where the user can 'plan ahead' and personalise the tool to better fit his or her work habits and preferences and improve them with AI-powered recommendations to work smarter, not harder.

The data-driven feedback provided by MyAnalytics is organised in four categories corresponding to the analytics performed at the backend of the system: Focus time, wellbeing, network and collaboration, which will be detailed further in the next section. Through these categories, MyAnalytics communicates what it considers ideal work behaviour: being structured and good at managing your time, getting things done in designated time slots, collaborating with others – but not too much, and not being digitally available for work all the time. The system calls for the employee to take responsibility in optimising both matters of personal productivity and wellbeing, calling on a moral imperative to become the best possible version of their working self, as has also been argued in other self-tracking research (Gregg, 2018; Kristensen, 2022; Till, 2019). Indeed, MyAnalytics presents itself not as an assistant (Winikoff et al., 2021) or even a coach but rather as a set of possibly well-meant *nudges* to regulate working behaviour towards desirable ends. Karen Yeung, building on the work of Thaler and Sunstein (2008), describes data-driven nudging as a digital decision-guidance process

> designed so that it is not the machine, but the targeted individual, who makes the relevant decision. These technologies seek to direct or guide the individual's decision-making processes in ways identified by the underlying software algorithm as 'optimal', by offering 'suggestions' intended to prompt the user to make decisions preferred by the choice architect.
>
> *(Yeung, 2017: 121)*

What seems implied in Yeung's theoretical argument is a characteristic of data-driven nudging as a form of digital paternalism or soft power which cultivates a specific set of work behaviours as optimal, but without a determining force. Arguably, these optimal behaviours are mirrored in public discourses of disconnection and distraction, digital media as threats to time management, work/life balance and so on, and thus presumably also to some extent recognised by the users as relevant.

Communicating with MyAnalytics

MyAnalytics tries to address a perceived and widespread problem of modern work life: being 'pressed for time' (Wajcman, 2015). The system is open-ended, in the sense that it can accommodate different working styles and preferences. Users can customise the analytics and recommendations to make a better fit with the concrete realities of their work tasks and everyday lives. This might, for instance, not only include changing the hours of the regular workday from the eight to five default setting with implications for the statistics for silent days and thus MyAnalytics' assessment of overall wellbeing. But it could also include actively defining how much focus time one needs, and when this should be placed in the weekly calendar, or whose emails should be marked as important. Such customisations communicate the users' contextual preferences to MyAnalytics and change the system in minimal ways in what, over time, becomes a recursive feedback loop. Even if MyAnalytics showcases the paternalist tendencies of pushed and imposed forms of self-tracking, the user controls for customising the service (or turning its individual analytics off entirely) also testify to the agency of users in altering – even optimising – digital systems by way of more fine-grained data input.

Yeung (2017) speaks to this idea, even if she emphasises the power of the algo-rithm designer who acts on user input to reconfigure the system: Every change brought about by user input, however small, has implications beyond that indi-vidual user and the organisation at which she is employed to the nudging of users in other organisations operating with Office365. By extension: through engaging in customisation of MyAnalytics the individual becomes more actively involved in deciding what should be deemed optimal and desirable work behaviour. At the same time, it remains an open and empirical question of how users actually respond to and possibly act on the nudges provided by MyAnalytics.

As a technology for self-tracking at work, MyAnalytics is placed in an everyday context where individuals put media to use in order to get things done, balance the demands of various domains and their own desires and preferences to lead a good life with whatever means available to them. Recruiting informants, I have come across many people who are somewhat aware that MyAnalytics exists but have not bothered to engage with the statistics, visualisations and small nudges it offers for adjusting work habits. Indifference to the system is characteristic also of some of the people I interviewed for the research underpinning this chapter. But I have also encountered informants for whom MyAnalytics is a welcome intervention for reflecting on and possibly improving their work habits, and who speak fondly about how the insights the system delivers have helped them fine-tune specific aspects of their work habits and work ethics to find greater happiness and meaning in their work lives. Such empirical variations in self-tracking with MyAnalytics, from distancing oneself to purposively appropriating the system, speak to the need for contextual, everyday grounding of research on self-tracking at work in the empirical realities of users when assessing the possible impacts of such AI-enhancement of work practices. I will offer brief examples of users' engagement with each of the four insight categories of MyAnalytics later.

Focus time describes the degree to which the employee has time to complete work tasks, defined as time where the employee is not attending meetings or doing email. Insights are delivered in a simple diagram comparing the time spent on meetings with other activities. On the user dashboard, data can be further explored to find out if specific times of day or days a week indicate a particular pattern, for example days with meetings back-to-back. In addition to this, *Focus time* has a feature that urges the employee to give priority to undisturbed 'deep work' when planning ahead. If there are many meetings in the calendar, MyAnalytics will sug-gest you reserve slots for focusing. In fact, it offers to do this for you, so you cannot be booked for another meeting during this time. It can also be made to ensure that while in focus time slots, the employee is not disturbed by incoming emails, chat messages or calls. An informant who is an early-career mid-level manager in a law firm and often needs to guide her employees' work or clear the cases they produce has customised the system to secure daily focus time at a specific two-hour time slot and regulate her availability for co-workers. She further asserts that by simply bringing the need for focus time out in the open, MyAnalytics has helped her talk with her employees about how they would like to work together as a team while also giving each other some space. Other informants point to the issue that some weeks might be heavy with customer meetings and thus make the supposed ideal of securing focus time each day or week impossible and less desirable.

The *Wellbeing* category directs the user's attention to issues of work/life balance. The premise is that the user needs to log off from work to 'recharge the batteries' to work productively over time. This means having quiet days with little to no work-related activity (e.g. on weekends) and not engaging in excessive email activity after normal office hours. Insights on wellbeing are displayed in a figure and calendar mapping the time an employee has to recharge and are accompanied by an automatic notification reminding the employee that he or she should not work after hours. None of the informants in the study find wellbeing a particularly helpful insight or nudge: one informant, whose job in municipal government includes evening meetings with a political board, laconically remarks that insisting on silent time after normal hours (defined by the system as after 5 p.m.) is not possible in her job. In her everyday life, working and family life blend flexibly according to her work schedule, and compensating for late office hours, she will often take afternoons off to be with her family. The lack of interest in the category might also be related to the empirical context in question. Collective worker rights in the Danish welfare state and legally required regular workplace assessments in a highly regulated job market may already bolster employees against excessive overwork. Hence, MyAnalytics is not needed to regulate work time.

The *Network* category displays in a list and network visualisation, which people are your most frequent – and presumably most important – interaction partners at work as gauged from existing communicative exchanges in Office365. This insight can then further be used to suggest that email from these important others is flagged to receive more immediate and prominent attention in the employee's inbox. On top of this, MyAnalytics will automatically generate suggestions for people the employee might want to reconnect with. It thus suggests that networking is an important skill – one that the employee should seek to groom. Only one informant, a project manager who coordinates projects with many internal and external stakeholders at an IT company, speaks of having used this feature at all. The quick overview provided by MyAnalytics functions as a checklist to see if she is on top of communication with her stakeholders.

Finally, *Collaboration* shows how much time the user spends on actively collaborating with others in meetings; it also displays when meetings are booked (on the fly or several days ahead) and notes if the employee has many meetings with an overlap in the people attending. Based on this insight, MyAnalytics will notify the employee if there appears to be high meeting activity with specific collaborators and ask the user to consider whether something can be done to optimise meeting activity, for instance, whether attendance of several people from the employee's division is necessary or some could be spared to attend a meeting. Furthermore, MyAnalytics correlates email activity and the like with the time slots where the employee is marked as in meetings. In this way, the insights seek to make the user aware of presumably inappropriate behaviour, such as doing your email while in a meeting. One informant, a middle-aged developer in a software company where MyAnalytics has been actively pursued for some time, asserts that while it may take some time to get used to it, 'the insight that it provides is actually useful . . . for instance the thing about writing emails while in a meeting, I have thought about it, and yes, I agree it is a bad habit'. MyAnalytics has pointed out something obvious to him and has actually made him change his meeting behaviour, by igniting reflections about work ethics; specifically, how to be a good co-worker.

These empirical examples certainly do not exhaust the analysis of the embedding of MyAnalytics in everyday life. But they do testify to how employees may make a critical and selective engagement with the nudges it provides, because these do not always fit the practical realities of work.

For some of my informants, MyAnalytics appears completely unnecessary: it responds to a problem they do not recognise. An informant, a middle manager in public administration, whose working style is already structured systematically in a way that 'works just fine' does not feel any need to cultivate better habits and does not find anything interesting or surprising in the MyAnalytics insights; hence, opening her weekly emails or the dashboard is perceived as a waste of time. Ironically, as this example might suggest, MyAnalytics becomes part of the problem (excess email, time wastefulness at work) that it tries to address. Another informant describes how MyAnalytics only covers some of her work; she receives MyAnalytics in the context of her part-time employment in an international company. Yet, her time management challenge is to juggle several part-time jobs while studying for an academic degree, and since MyAnalytics only covers one out of three of the professional contexts in which her working life unfolds, it is irrelevant for her, even if she likes the idea of the system.

As such examples suggest, pushed self-tracking may simply be ignored by users. Distancing themselves may be conceived as a resistant practice to suggest that technological visions do not always travel well and meaningfully into the everyday realities of people. One important addition to this observation is that by being a default add-on in Office365, MyAnalytics is in fact implemented and imposed by Microsoft and not by the organisations for which my informants work. Without an explicit push from organisational management, and by extension without an explicitly communicated organisational interest in MyAnalytics, it may be easy to ignore new technologies for self-tracking at work. MyAnalytics is a technological instrument for normative control over work performance through homogenisation of values, attitudes and behaviours, but if workers and organisations do not really pay attention to it, we should be careful and critical when assessing if and how its paternalist nudges become practically meaningful. In that sense, the analysis also speaks to the need to look beyond technological fixes when addressing structural problems of modern work.

AI-enhanced self-tracking for a good work life?

MyAnalytics presents an AI-powered solution to address something that is arguably experienced by white-collar workers as an everyday problem, even if a luxury one: managing one's time and struggling to juggle the various demands and duties emanating from both personal everyday life and working life. The brief empirical examples presented as part of my communication-based analyses align with a recent qualitative study of MyAnalytics (Winikoff et al., 2021) but should not be read as generalising statements over the use of MyAnalytics for smart work. However, they do highlight crucial elements that we should not leave unaccounted for when assessing the consequences of self-tracking as everyday AI.

One, echoing self-tracking research in other domains than work: self-tracking may be pursued for multiple ends, only some of which may align closely with those anticipated by technological design (e.g. Ruckenstein, 2014; Chung et al., 2017). As Kristensen (2022: 4) summarises, we should recognise the 'co-evolving of technology and the self to produce new experiences, emotions, and transformations of the self, thereby actively responding to and even resisting societal values of productivity and efficiency'. Two, context is key to understanding the role that MyAnalytics might play in the empirical realities of employees – and a good and meaningful work life might look different from that communicated by the system, depending on the employee, the job description and the organisation in question. While technological solutions of everyday AI at work might push for standardisation or homogenisation of work values and ethics under a broad logic of optimisation, users are not just prey for automated communication; they actively contribute to the continuous shaping and appropriation of digital systems to their own ends. Furthermore, since cultural contexts and values of work vary significantly, we cannot assume that prospects of corporate control through data are exercised, appropriated or resisted uniformly. Digital applications for self-tracking of wellness and productivity developed in an American context may be differently received and regulated in a Nordic welfare state like Denmark. A communicative perspective reminds us that even under regimes of pushed and imposed self-tracking, users' agentic capabilities for meaning-making and their micro-adjustments of systems through communicative practice contribute to shaping the futures of digital and datafied work. In advancing our understanding of automated communication and everyday AI, we should strive for empirical, contextually grounded research that can help detail not only what tracking and datafication does to people but also how people (and organisations) communicate through and with data, if at all, in managing their time and striving towards good working lives.

References

Ajunwa I, Crawford K and Schultz J (2016) Limitless Worker Surveillance. *California Law Review* 735.

Andrejevic M (2020) *Automated Media*. London and New York: Routledge.

Ball K (2010) Workplace Surveillance: An Overview. *Labor History* 51: 87–106.

Calvard T (2019) Integrating Social Scientific Perspectives on the Quantified Employee Self. *Social Sciences* 8.

Carey JW (1992 [1989]) *Communication as Culture. Essays on Media and Society*. London and New York: Routledge.

Chung CF, Jensen NG, ShklovskI I and Munson S (2017) Finding the Right Fit: Understanding Health Tracking in Workplace Wellness Programs. In: *CHI'17. CHI Conference on Human Factors in Computing Systems*, Denver, US 6. 11 May. Association for Computing Machinery, ACM Annual Conference on Human Factors in Computing Systems (CHI), 4875–86.

Crawford K (2021) *Atlas of AI. Power, Politics and the Planetary Costs of Artificial Intelligence.* New Haven: Yale University Press.

Delfanti A and Frey B (2021) Humanly Extended Automation or the Future of Work Seen through Amazon Patents. *Science, Technology, & Human Values* 46: 655–82.

Elmholdt KT, Elmholdt C and Haarh L (2021) Counting Sleep: Ambiguity, Aspirational Control and the Politics of Digital Self-tracking at Work. *Organization Science* 28: 164–85.

Eubanks V (2018) *Automating Inequality. How High-Tech Tools Profile, Police and Punish the Poor.* New York: St. Martin's Press.

Gregg M (2018) *Counterproductive. Time Management in the Knowledge Economy.* Durham: Duke University Press.

Guyard C and Kaun A (2018) Workfulness: Governing the Disobedient Brain. *Journal of Cultural Economy* 11: 535–48.

Hull G and Pasquale F (2019) Toward a Critical Theory of Corporate Wellness. *BioSocieties* 13: 190–212.

Kitchin R (2014) *The Data Revolution: Big Data, Open Data, Data Infrastructures and Their Consequences.* London: Sage.

Kristensen DB (2022) The Optimised and Enhanced Self. Experiences of the Self and the Making of Societal Values. In: Bruun MH, Wahlberg A, Douglas-Jones R, Hasse C, Høyer KL, Kristensen DB and Winthereik BR (eds) *Handbook of the Anthropology of Technology.* London: Palgrave, NP.

Lomborg S and Frandsen K (2016) Self-tracking as Communication. *Information, Communication & Society* 9: 1015–27.

Lupton D (1995) *The Imperative of Health: Public Health and the Regulated Body.* London and New York: Sage.

Lupton D (2016) The Diverse Domains of Quantified Selves: Self-tracking Modes and Dataveillance. *Economy and Society* 45: 101–22.

Moore PV (2018) *The Quantified Self in Precarity: Work, Technology, and What Counts.* London and New York: Routledge.

Moore PV, Upchurch M and Whittaker X (2017) Introduction. In: Moore PV, Upchurch M and Whittaker X (eds) *Humans and Machines at Work: Monitoring, Surveillance and Automation in Contemporary Capitalism.* London: Palgrave.

Newport C (2016) *Deep Work. Rules for Focused Success in a Distracted World.* New York: Grand Central Publishing.

Pink S, Sumartojo E, Lupton D and Hayes Labond C (2017) Mundane Data: The Routines, Contingencies and Accomplishments of Digital Living. *Big Data & Society* 2017: 1–12.

Rosenblat A and Stark L (2016) Algorithmic Labor and Information Asymmetries: A Case Study of Uber's Drivers. *International Journal of Communication* 10: 3758–84.

Ruckenstein M (2014) Visualized and Interacted Life: Personal Analytics and Engagements with Data Doubles. *Societies* 4: 68–84.

Sánchez-Monedero J and Dencik L (2019) The Datafication of the Workplace. *Working Paper of the Data Justice Project.* Cardiff University, Cardiff. Available at: https://datajusticeproject.net/wp-content/uploads/sites/30/2019/05/Report-The-datafication-of-the-workplace.pdf

Spataro J (2020) Our Commitment to Privacy in Microsoft Productivity Score. Available at: www.microsoft.com/en-us/microsoft-365/blog/2020/12/01/our-commitment-to-privacy-in-microsoft-productivity-score/?ranMID=24542&ranEAID=nOD/rLJHOac&ranSiteID=nOD_rLJHOac-lgi_FJtapLqGM3KOcRbW1Q&epi=nOD_rLJHOac-lgi_FJtapLqGM3KOcRbW1Q&irgwc=1&OCID=AID2000142_aff_7593_1243925&tduid=%28ir__1v1s6at0dckfq3h1xka03fe3c22xuzkcttjrcwwv00%29%287593

%29%281243925%29%28nOD_rLJHOac-lgi_FJtapLqGM3KOcRbW1Q%29%28%29 &irclickid=_1v1s6at0dckfq3h1xka03fe3c22xuzkcttjrcwwv00 (accessed 29 April 2021).

Thaler RH and Sunstein CR (2008) *Nudge: Improving Decisions about Health, Wealth, and Happiness.* New Haven, CT: Yale University Press.

Till C (2019) Creating 'Automatic Subjects': Corporate Wellness and Self-tracking. *Health Sociology Review* 23: 418–35.

Wajcman J (2015) *Pressed for Time: The Acceleration of Life in Digital Capitalism.* Chicago, IL: University of Chicago Press.

Winikoff M, Cranefield J, Li J, Richter A and Doyle C (2021) The Advent of Digital Productivity Assistants: The Case of Microsoft My Analytics. In: *The 54th Hawaii International Conference on System Science.* Hawaii, 5–8 January. INSNA.

Yeung K (2017) 'Hypernudge': Big Data as a Mode of Regulation by Design. *Information, Communication & Society* 20: 118–36.

9

EXPLORING ADM IN CLINICAL DECISION-MAKING

Healthcare experts encountering digital automation

Magnus Bergquist and Bertil Rolandsson

Introduction

> Based on either deep learning or machine learning, AI can be defined as learning by looking at a sufficient number of people and combining the various elements of the desired data to achieve the desired results. This makes it possible to point out the probability of a certain disease, tumour, a progression of MS or progression of lung metastasis, for example. It can be good to gain this information, but it is important that a radiologist can raise questions and ask whether the suggested result would be reasonable. Is the result correct? You can only do this by looking closely yourself!
>
> (Chief Physician in Radiology)

The claim by some researchers that automated decision-making (ADM) is about to outperform healthcare experts and make the need for clinical reasoning dispensable (e.g. Susskind and Susskind, 2015) misses the important role of exploration and continuous knowledge development in healthcare practices. This was clear to the chief physician quoted earlier, when she stated that ADM needs human expertise to provide the level of support needed in clinical decision-making. From her perspective as a clinical expert, ADM is a technology that, similar to many previous healthcare technologies, continues to demand human expertise when the task is to diagnose patients based on clinical data. Therefore, she argued that healthcare professionals must be involved in exploring and developing the technology to enhance their understanding of how to apply expertise in a setting based on reasoning, diagnosis and priorities. In doing so, healthcare experts actively take part in creating the context for how ADM is to be understood and used in clinical decision-making.

The chief physician's reasoning contradicts some recent research and public debate arguing that ADM as a technology is capable of reducing the importance of

DOI: 10.4324/9781003170884-12

humans in professional decision-making and making the role of human expertise less needed (Lester, 2020; Shilo et al., 2020). Although that certainly is true in some cases, it is obvious that the complexity of contextual contingencies illustrated in the quote challenges us to develop our understanding of how decision-making actually happens in practice, and how different institutional structures are drawn into situations where technologies for decision-making are used. An institutional approach to studying ADM involves a sensitivity to how practices and contextual variations shape decision-making (Nicolini, 2011).

In this chapter, we are therefore interested in how the delegation of decision-making (not the decision in itself) to data-driven, algorithmically controlled systems plays out (cf. AlgorithmWatch, 2019). On the one hand, we identify how ADM is linked to several forms of creativity, shaping the exploration and testing of new possibilities to improve care. On the other hand, this creativity is dependent on context-specific contingencies and strains due to institutionalised demands for accountability that are part of professional identity in healthcare work.

The fieldwork underpinning the chapter focused on ongoing projects at two hospitals in West Sweden where healthcare experts (doctors, nurses, clinical specialists, physicists, psychologists and engineers) were involved in developing ADM for clinical decision-making. They did this not only because they were convinced that ADM could be of great help in clinical decision-making but also because they wanted to be accountable for decisions that involve support from ADM technologies. These healthcare workers approached ADM by actively designing and integrating the technology into their own decision-making practices.

The analysis is based on semi-structured interviews with 20 healthcare experts active at these two hospitals. One hospital was a large university facility and the other was a smaller regional hospital. The interview participants represent different clinical specialisations engaged in a variety of ADM projects centred around developing and using artificial intelligence (AI) and machine learning to diagnose and predict illnesses in risk groups or provide early warning systems in mental health care or emergency care. Some of these projects make use of combinations of retrospective and real-time health data that could support evidence-based decision-making, individualised care and precision medicine (Ashfaq et al., 2019; Blom et al., 2019; Øvrelid et al., 2019). The majority of our interviewees represented radiology, which is described as one of the most ADM-intensive domains in healthcare (Susskind and Susskind, 2015) because of the potential for algorithms to visualise and analyse segments in radiology images (e.g. brain lesion segmentation). We also interviewed engineers involved in developing ADM solutions, together with clinical experts.

Digitisation of healthcare work and discretionary decision-making

The development of ADM in healthcare is a response to challenges created by recent digitisation where analogue healthcare data are increasingly digitised, thereby creating opportunities for innovation and new modes of clinical work, dependent on increasing amounts of data and moving into new application areas (Tresp et al., 2016). This type of digitisation of clinical health data has had two major consequences for clinicians. First, the sheer volume of information that needs to be processed for every case has multiplied and is escalating exponentially. Second, the granularity of the provided information increases. For example, high-resolution Computerised Tomographic (CT) radiology generates several hundred images per examination (Gryska et al., 2021; cf. Raghupathi and Raghupathi, 2014). ADM is introduced as an approach to extract valuable information from this growing amount of data to improve the basis for decision-making.

While this type of technological innovation bodes for substantial changes (Baumol et al., 2012), healthcare experts encounter such changes by bearing in mind demands for different quality criteria (Dent et al., 2016) that require their professionalism and awareness of regulations and accountability (Molander and Grimen, 2010). Guided by experts' experiences and proficiency in tackling ambiguities founded on responsibilities and practices protected and preserved by the profession, decision-making in healthcare is shaped by their *discretion*, providing judgements that comply with regulations and professional practices. Healthcare experts engaging with ADM are thereby also constantly involved in translating knowledge and standards to suit the needs and features of the case in hand and are held accountable for the decisions made and how discretion is practised (Noordegraaf, 2020). In this chapter, we focus on this interplay between the development of ADM and the enactment of healthcare professionals' discretionary judgements.

These types of discretionary judgements can be divided into *procedural discretion*, responding to demands of administrative fairness (e.g. involving the registration and storing of citizens' data), or *substantial discretion*, responding to demands of accuracy (e.g. expert judgements on clinical matters and normative content) (Sainsbury, 2001; Feldman, 2001). Previous research has focused on ADM in the context of so-called street-level bureaucracy, primarily investigating procedural discretion among officials using digital technologies to store and handle client data and the challenges of delegating what is defined as routine decisions to ADM technology (Busch and Henriksen, 2018; Bullock, 2019). Technological solutions in these studies often appear to underpin 'bureaucratic managerialism' in a way that undermines officials' discretionary autonomy (Jorna and Wagenaar, 2007; Petrakaki et al., 2016). Digital technologies force them to comply with organisational routines and formalised relationships with citizens (Buffat, 2015).

Less is known about the healthcare experts, as studies investigating their discretionary use of ADM are rare. In the context of ADM, healthcare experts emerge as professionals primarily encountering demands for substantial discretion, requiring their ability to independently tackle ambiguities in line with state-of-the-art knowledge and demands for accuracy in decision-making (Molander and Grimen, 2010; Noordegraaf, 2020). The fact that they draw on substantial discretion and their ability to consider how to translate their expertise to the needs and features of the case in hand indicates that their own professional experience and ability to creatively identify and use data through improvisation play a crucial role in decision-making. Whenever healthcare experts aim to reach a decision, it is part of their job to explore all varieties of associations and potential outcomes that can be identified to make sure that every aspect of the problem in hand is scrutinised.

Discretionary decision-making in this type of exploratory practice demands professional autonomy, but we have to recognise that, at the same time, healthcare experts are also conditioned by rules, regulations and ethical considerations (Molander and Grimen, 2010). The way they explore the space for discretion reflects available standards, regulations and organisational constraints that constitute the context for their professional practice. They relate to and thereby act upon different demands for accountability, considering the consequences of not delivering on goals, violating codes of conduct or failing to comply with available frames of reference (Feldman, 2001; Sainsbury, 2001).

Experts exploring their space for expertise

In contrast to much of the literature on ADM that describes the anxieties connected to its use and how it might reduce work autonomy, outperform healthcare experts and make them dispensable (Susskind and Susskind, 2015), the participants in our study were generally very engaged in expectations connected to digital automation technologies. They experienced ADM as the next step in the constant exploration and testing of new opportunities to improve care and thus push the boundaries for what can be known and accomplished. The background for this was related to previous development by engineers and clinicians to enable new technological possibilities. For example, the radiologists we interviewed had a long history of introducing new modalities such as computer tomography and magnetic resonance cameras that allowed them to image the inside of the body in new ways and thus explore new territories for diagnostic opportunities. This development had resulted in both increased volumes of images and more complex diagnostic scenarios. To meet challenges in a growing inflow of patients in combination with technological possibilities that drastically expanded the possible types of examinations, the clinicians needed support in their daily practice.

Although the experts we interviewed confirmed that new AI technologies would be likely to change the way they worked, their accounts of continuous demands for discretionary capabilities moved beyond suggested assumptions about a strained relationship between automated and human decisions. A more complex relationship between ADM and healthcare experts emerged where discretionary capabilities are reshaped in interactions with the new technology. Therefore, ADM is seen as part of a broader assemblage of technologies and activities involving experts who judiciously explore and critically test new technologies. The interviewees depict a practice of continuous exploration in which ADM both supports and challenges demands for professional discretion, including rules and procedures for how to be accountable in decision-making. In the following section, we describe how ADM became part of this *practice of exploration* and how discretion within this practice shaped the interaction with ADM: first, in patient-related diagnostic work, and second, in the *organisation of discretion* where the experts sort out how to administer rapidly increasing amounts of data.

Exploring and deepening professional experience

A recurring theme in the interviews is how clinical experts see opportunities to deepen their expertise on diagnostic work by using ADM to sort out clinical evidence in the increasing stream of data. The radiologists described how automated segmentation – outlining visceral malign areas on a radiology image – could be a way of automatically identifying the progression or reduction of tumours and other forms of bone, tissue and organ injuries. Measuring the volume of tumours with various shapes is not only difficult and time-consuming but also, to increase the quality of the diagnosis, radiologists need to combine image data with other data:

> We need to draw it [on the image] and this takes time. Drawing the gland is something that AI can do; it's something we believe can be a relatively simple task for the computer and the type of task that I believe AI will be highly useful in accomplishing. . . . Furthermore, we have data from the pathologists about what these MR-images actually depicted and we can use this to program the algorithm.

Several initiatives are mentioned to connect previously unconnected datasets from different parts of the healthcare process, such as lab tests, pathology, radio genomics, EHR and medication lists, to push the analytical capacity forward. By combining different types of datasets to reach the next level of diagnosis, it is possible to explore the scope for achievement. In the specific case of radiology, the emerging practice of exploring image data with AI depends on the continuous improvement of image data quality from modalities such as magnetic resonance imaging (MRI) scanners. It also depends on which methods to use when analysing hundreds of images of cross-sectional slices of organs generated from a single computed tomography (CT) examination. The paradox highlighted by the experts is that the huge data volumes generated by technology raise expectations on discretionary capabilities that need automation to cope with the vast amounts of data to reassure diagnostic quality. The better the experts who participated in our study become at using ADM for decision-making, the more dependent they will become on the algorithms and thus the need to improve them. Exploration becomes a necessity.

Exploration in the context of accountability

Healthcare professionals' decision-making builds on discretion, which is the ability to make good and sound decisions based on evidence and expertise. When the experts we interviewed reflected on technical development, they expressed concerns that it is becoming more difficult to make evidence-based decisions as a result of the vast increase of empirical data. They argued that it was easier to make decisions in the past because of a smaller series of manually collected data with fewer samples containing less detail. When faced with increasingly bigger and digitally integrated datasets, the demands for improved accuracy in decision-making become an urgent matter. A radiologist explained that as professionals, they face a situation where it is increasingly difficult to be accountable for considering all evidence and all probable findings in available data:

> Making decisions based on radiology images is a complex task. It's not simple but very difficult. Not only is it difficult to interpret an image from a magnet resonance camera, the actual proper use of the camera is a challenge that requires much education. Therefore, we need tools, both to help us to master the increasing volume of radiology images as a result of new camera methods and also to help us with the diagnostic work.

The radiologist quoted earlier argued that getting involved in developing ADM can be a way to develop knowledge and competence for how to exercise judgement. ADM must be designed to allow for data exploration, for hypothesis testing and for trying out technological potential in interaction with ADM to identify the best tools for responsible decision-making. Therefore, to become a skilful and competent user of ADM, it is important to engage with the technology as a means that requires both creativity and the ability to associate data sources with each other in an evaluative and critical manner. A senior radiologist and researcher who defines herself as belonging to 'the older generation' argued that the clinician has a moral obligation to be explorative and go beyond pre-given tasks. The tasks include the formal request from the referring physician to explore every possible aspect of the radiology image in combination with other data that could be pieces in the diagnostics puzzle. However, new imaging possibilities have generated expectations from referring physicians that radiologists can give well-grounded answers, including when it comes to wide and blurry questions. Radiologists need support in diagnostic work in order to make good decisions in this new situation of possibilities and demands. Therefore, they assume that human experts are most likely to continue to be accountable in clinical decision-making in the future. Humans have the ability to associatively connect seemingly unrelated indications to identify an illness or to be able to write off a suspected diagnosis in a way that a programmed decision support system cannot achieve.

Quality assurance of judgement in exploration

To provide reliable interpretations and diagnoses, healthcare experts have to be skilled and actively engage with the technology. Familiarity with the specifics of existing technologies (modalities, software, image manipulation tools, etc.) is a precondition for the ability to provide substantial and accurate decision-making. Involvement in developing ADM is a way to create the new skills needed to understand how ADM may change work practices or to provide a better understanding of how different brands of the same clinical system affect AI-based analyses. One of the engineers we interviewed was involved in developing ADM for radiologists and stressed the importance of interactions between experienced clinical and technical experts in designing ADM. This has the aim of ensuring that all aspects of the ADM solution are questioned and critically investigated to avoid the identified risk that ADM could make the interface between the expert and the technology opaque. The engineer explains:

> If I introduce a picture of Mickey Mouse, the computer will almost certainly identify some areas as cancer, whereas a human knows that something is wrong. If a hip prosthesis appears on a CT-image and the computer has never seen a hip prosthesis, we'll receive results that are completely wrong, while a human knows what a hip prosthesis is. This does not mean that the computer is stupid, but that it just hasn't been trained to assimilate enough data. However, somehow, humans have been. In the end, this means that we'll have to wait quite some time before doctors don't have to look at the pictures, but we already know that their accuracy improves when computers look at the pictures first.

Since ADM relies on machine learning based on provided datasets, it is clear to the interviewees that the technology is no more intelligent than it is trained to be. As a consequence, technology cannot be held accountable for whatever diagnosis and treatment it suggests. Since healthcare experts continue to be accountable in ADM-supported decision-making, they feel a need to develop new skills based on an understanding of machine learning technology to judge the quality of the machine's output. This new ability to critically interpret automatically generated findings is seen as crucial to achieve accuracy in each specific case and to understand how they can be accountable in this new human–machine joint learning situation. The demand for critical expertise in human–machine learning also recurs among other experts. A psychologist we interviewed, who was involved in developing an AI-based 'early warning system' for identifying depression among teenagers, argued that it is crucial to tackle the risk of misjudging automatically generated warnings. In particular, they have to avoid identifying individuals as mentally ill too early and thereby turning people into patients without real reasons.

> We have to avoid negative consequences [of predictions] because we can't diagnose anyone one year in advance. Moreover, if we look at false positives, that is, those who are identified by an AI as being sick without being it, can be exposed to unnecessary treatments. Another group comprises those who display all the characteristics we should act upon, including anxiety and depression, without being identified. There are several categories to consider, which makes it important to identify the accurate cases for a certain level of treatment. Otherwise, it would be unethical.

The interviewed psychologist underscores that machine learning predictions based on historical data can result in false conclusions and turn people into patients instead of curing them. For him, it was essential that he could reach a reasonable judgement to be accountable for decisions based on a substantial and accurate analysis of existing data. Algorithm-based decision-making highlights the role of accountability, not because 'the responsibility for decisions still lies with the humans who commission, develop and approve ADM systems' (Matzat, 2019: 4) but because accountability is an integrated part of healthcare professionals' ethics and work practice.

Organising the practice of exploration

Individual clinical expert decision-making – often as part of a diagnosis or a treatment plan – is part of the ongoing, everyday discretion that is challenged by ADM-supported explorations, including how to deal with opaque algorithms, context dependency, integrity issues, dependency on training data or hardware features. However, the daily activities of testing, trying out and exploring are also connected to different demands of organising the practice of exploration shaped by institutional requirements. Daily activities are concerned with issues such as how to use resources in an efficient way or how to prioritise cases when resources are limited. This connects ADM to institutional procedures and the role of peers in clinical decision-making. Organisational transparency and traceability make decision-making legitimate. Documentation, follow-ups and prioritisation with integrity are examples of practices that give decisions institutional legitimacy. For example, the sorting and evaluation of multiple sources of information are expected to be treated so that a decision can be traced back to its sources and clinical judgement can be evaluated.

In this context, ADM emerges both as a 'butler' that monitors activities based on large and complex datasets to acquire accuracy in diagnosing when this would otherwise require substantial human resources and as an advanced analytical 'colleague' supporting exploration: for instance, by suggesting evidence-based diagnoses and recommendations. While the demand for accountability means that in the end, the machine cannot be responsible for the actual decision-making, the interviewees' appreciation of ADM as a 'butler' depends on its capacity to tackle escalating data volumes. ADM is an increasingly important tool for scanning, structuring, prioritising and evaluating clinical data in daily practice. One of the most sought-after solutions proposed in the interviews was an algorithm that can define 'normal' cases and remove them from the diagnostic process:

> It would really make it easier if we had an image assessment tool for radiology that could sort cases into 'healthy-healthy-sick-healthy-healthy'. It would make it much easier if we could avoid examining cases that are classified 'with no remarks' or classified as 'with no major remarks' and concentrate on the patients who really are ill.

As a butler, ADM must be able to act in accordance with predefined directives, no matter how much data are processed, so that it can facilitate the experts' exploration and competency development. The example outlined earlier illustrates how this could be the case when ADM identifies patients that demand resources. The clinical experts described how ADM can help them understand the wider context for decision-making by comparing patients to other patients in the same cohort or to personalise care by validating contextually informed decisions and identifying any possible bias as a way to promote accurate and efficient care to patients.

In line with such a demand for evidence, the clinicians also raised concerns that some of the new ADM-technologies were poorly tested responses to trends, depicting them as quick fixes to complex problems. This critical approach, however, is not primarily an expression of distrust in technology but deeply rooted in the need to always be accountable. As part of their constant exploration, decisions need to be accurate both as an expression of the clinician's individual expertise and as an expression of the clinician as a representative for an organisation that needs to ensure patients have fair treatment in accordance with an agreed evidence-based standard. Exploring data that can lead to findings and improve clinicians' conditions for reaching accurate judgements is linked to arguments about awareness of procedural quality and administrative fairness (Feldman, 2001). One of the chief physicians explains:

> Sometimes, I meet colleagues who talk about a system they bought that they believe is great because it can spot small changes in the lungs and give a quick answer. But is that the correct answer? How do we know if it is? Often, they don't really know. Just because it's available on the market doesn't mean that it's certified from a diagnostic perspective. It might work under some conditions but not others. For instance, the result can change depending on the brand of the machine, the way an image is taken, the particular patient group, age or something else and we don't know what's related to what. We might need one AI system for one type of questions and another system for another type of questions. The best is to use test data that you are familiar with and compare the machine's answer to that created by a human.

The chief physician quoted earlier states that introducing ADM as a technological system into the decision-making process can risk obscuring the clinician's responsibility towards peers and the profession. Several of the clinicians we interviewed highlighted the importance of ensuring that they take professional and moral responsibility in decision-making. However, ADM involves a risk that decisions are exposed to what has been named 'agency laundering' (Rubel et al., 2019): namely the delegation to an AI of responsibility for the actual decision. When ADM acts upon data, it can obfuscate the moral responsibility of the clinician to be accountable for the decision.

The fact that the clinicians' professionalism always has to be the last outpost to ensure that decisions are made with responsibility makes the potential obfuscation of agency problematic. It is important to ensure that a clinician is the last entity in the chain to analyse the image and ultimately shoulders the responsibility for the final decision. A radiologist at the beginning of his career, who collaborated with engineers in different research projects on the opportunities of AI as a diagnostic tool, reflected that there are few options. Companies developing healthcare systems based on ADM will probably not take responsibility for errors that could occur. Therefore, what has been referred to here as agency laundering will be an emerging challenge for the profession.

Generalisation and personalisation

Another theme that connects ADM to the organisational aspects of decision-making is the idea that care must be balanced in relation to formal demands for fairness. Those patients who are in most need of care should be prioritised, which also means that a doctor should avoid further treatment if it is not likely that the patient will recover or make good use of the treatment. The experts are held accountable not only for how well the individual patient is treated but also for how well decisions are based on optimising shared resources between patients.

To achieve this, healthcare professionals have developed a system of generalisation and personalisation. Generalisation involves using large-scale data to identify risk groups, illnesses that could be traced with DNA, or personal traits and lifestyle issues crucial to attempts to prevent or minimise illness in specific patient cohorts. Personalisation is the ability to target individual needs using precision medicine and by developing a better understanding of the differences between patient groups, gender, age, socioeconomic differences, etc. From the perspective of ADM, there are opportunities to use algorithms to generalise particularities and see the individual case in a wider context, as well as use these generalisations to personalise care and provide each patient with care grounded in the specific individual's medical and personal needs.

Both personalisation and generalisation require an automatic analysis of large integrated datasets to identify patterns and to see how these patterns are dependent on various variables. One clinician, active as Chief Physician, explains his expectations on such a development:

> The next step might be to integrate data from lab tests, clinical data, genetic tests, age and magnet resonance camera data, then create a risk profile for this specific individual patient.

To generalise and personalise care, more data nonetheless must be identified and integrated into databases. This poses both technical and ethical questions to the specialists developing ADM. Integrity issues must be resolved and different kinds of data that can otherwise be difficult to compare must be standardised and integrated into general systems. When discussing ADM, some of the clinicians we interviewed raised the question of how the technology can support evidence-based decisions: for example, to deny a patient a specific treatment based on big data analysis of the outcome of that particular treatment. With ADM, the doctor would potentially have more detailed insights into the impact of that particular treatment in relation to age, social strata, previous treatments, gender, home address and so on. With an increasing stream of patients and new knowledge about the impact of treatment in relation to other statistically significant data points, it would be possible for the doctor to effectively administer patients in a fair manner while being accountable according to the norms and standards when it comes to their actions. An emergency ward doctor argued that the liability in exploring different opportunities for fair decisions always comes with a need for accountability. ADM could help the doctor remain accountable when making difficult decisions.

Therefore, for many of the healthcare staff who participated in our research, there were two ways in which ADM enabled them to practise in an exploratory mode. On the one hand, ADM creates opportunities to explore individuals and thus personalise their treatment: each patient should be able to have treatment tailored to their particular biological, physical, social and individual needs. On the other hand, big data in combination with machine learning create opportunities to investigate and identify high-level patterns and causality that could be used to identify and predict health issues among different population groups or for evidence-based standardised treatment plans for specific diagnosis: ADM enables quality certification for hospitals.

Conclusions

In this chapter, we have explored how healthcare experts' discretionary work and ADM become interwoven when applied in actual healthcare practices. The way algorithms are approached in different experimental research projects is quite different from how ADM has been implemented into the work of administrative staff described in previous studies (Ranerup and Henriksen, 2020) where ADM technologies are used for enhancing efficiency, effectiveness and standardisation, sometimes by replacing street-level bureaucrats (Bullock, 2019). In contrast, the healthcare experts we interviewed see ADM as a new tool that is integrated into an existing *practice of exploration* that has been an essential part of healthcare experts practice for decades.

The role of discretion is crucial to healthcare professionals' understanding of ADM as a tool and of themselves as experts promoting technology development. As experts, they do see themselves not merely as users of ADM but also as actors actively designing the case for use, seeking to understand how they could benefit from automated decision support. In highlighting this type of agency, our research has revealed that ADM remains an ambiguous tool, enabling professionals to manage complexity, as well as create conditions for an even more complex expert practice.

What stands out is the contradictory relationship between data and decision-making and how this tense relationship shapes discretion in the face of ADM. For clinical experts, discretion is founded on judgement and the ability to build decision-making on evidence that is compiled and analysed. The logic is that the more data available, the better the evidence, which results in well-founded clinical decisions. From that perspective, the participants in our research welcome the digitisation of clinical data such as health records, lab reports and radiology images. The more data they have access to, the more accountable they are: both in relation to norms and legal arrangements in the profession and in relation to their ability to meet expectations from patients, the healthcare organisation and broader public opinion (Roberts, 1991; Rolandsson, 2019). Simultaneously, their need to be accountable is challenged by the overwhelming access to digitised clinical data. It has become increasingly difficult to consider all available data in every case, and thus, it is more difficult to be accountable in relation to available evidence. Still, this context is required to make sense of how ADM is called for in clinical practice.

By recognising this context, we may understand how clinical experts integrate ADM into a practice of exploration. They use ADM as a means not only to consolidate evidence-based decision-making but also to push the limits of what can be known and mastered within the demands for substantial discretion (Molander and Grimen, 2010; Noordegraaf, 2020). In that sense, ADM is no different from any other new technology that has become available in healthcare. The application of ADM emerges as a continuation of previous exploratory practices enacted by healthcare experts known for seeking solutions to problems and innovating ways (Baldwin and von Hippel, 2011) to improve discretionary judgement.

Nevertheless, ADM emerges as being opaque and, therefore, difficult to grasp. This has consequences for discretion as it leads to reflections on how to ensure the relevance of the ADM proposed. In many instances, accordingly, the integration of this technology directs and shapes the practice of exploration by introducing negotiations for different possible explanations and interpretations, reflecting long-established practices that can be called upon to serve as sense-making devices for experts.

To be accountable in ADM-supported decision-making, healthcare experts develop new skills, based on their understanding of machine learning technology, which inform their expectations and evaluations of the performance of ADM. The ability to critically interpret automatically generated findings is seen as crucial for achieving accuracy in specific cases and for becoming accountable in this new human–machine joint learning situation. Thus, rather than outperforming the experts, the introduction of ADM raises new demands for experts and human accountability.

References

AlgorithmWatch (2019) Automating Society – Taking Stock of Automated Decision-Making in the EU. Available at: https://algorithmwatch.org/en/automating-society-2019/ (accessed 30 August 2021).

Ashfaq A, Sant'Anna A, LingMan M and Nowaczyk S (2019) Readmission Prediction Using Deep Learning on Electronic Health Records. *Journal of Biomedical Informatics* 97: 103256.

Baldwin C and von Hippel E (2011) Modeling a Paradigm Shift: From Producer Innovation to User and Open Collaborative Innovation. *Organization Science* 22(6): 1399–417.

Baumol WT, De Ferranti D, Malach M, Pablos-Méndez, Tabils HH and Wu LG (2012) *The Cost Disease. Why Computers Get Cheaper and Health Care Doesn't*. Yale: Yale University Press.

Blom MC, Ashfaq A, Sant'Anna A, Anderson PD and Lingman M (2019) Training Machine Learning Models to Predict 30-Day Mortality in Patients Discharged from the Emergency Department: A Retrospective, Population-based Registry Study. *BMJ Open* 9(8): e028015.

Buffat A (2015) Street-level Bureaucracy and e-Government. *Public Management Review* 17(1): 149–61.

Bullock JB (2019) Artificial Intelligence, Discretion and Bureaucracy. *American Review of Public Administration* 49(7): 751–61.

Busch PA and Henriksen HZ (2018) Digital Discretion – A Systematic Literature Review of ICT and Street-level-discretion. *Information Polity* 23(1): 3–28.

Dent M, Bourgeaul IL, Denis JL and Kuhlmann E (2016) General Introduction: The Changing World of Professions and Professionalism. In: Dent M, Bourgeaul IL, Denis JL and Kuhlmann (eds) *The Routledge Companion to the Professions and Professionalism*. New York: Routledge, 1–10.

Feldman M (2001) Social Limits to Discretion: An Organizational Perspective. In: Hawkins K (ed) *The Uses of Discretion*. Oxford: Oxford University Press, 163–84.

Gryska E, Schneiderman J, Björkman-Burtscher I and Heckemann R (2021) Automatic Brain Lesion Segmentation on Standard Magnetic Resonance Images: A Scoping Review. *BMJ Open* 11: e042660.

Jorna F and Wagenaar P (2007) The Iron Cage Strengthened? Discretion and Digital Discipline. *Public Administration* 85(1): 189–214.

Lester S (2020) New Technology and Professional Work. *Professions and Professionalism* 10(1): e3836.

Matzat L (2019) *Atlas of Automation – Automated Decision-making and Participation in Germany*. *AW AlgorithmWatch gGmbH*. Berlin: Bertelsmann. Available at: http://atlas-of-automation.net

Molander A and Grimen H (2010) Understanding Professional Discretion. In: Svensson LG and Evetts J (eds) *Sociology of Professions*. Göteborg: Bokförlaget Daidalos, 167–87.

Nicolini D (2011) Practice as the Site of Knowing: Insights from the Field of Telemedicine. *Organization Science* 22(3): 602–20.

Noordegraaf M (2020) Protective or Connective Professionalism? How Connected Professionals Can (Still) Act as Autonomous and Authoritative Experts. *Journal of Professions and Organization* 1(19): 205–23.

Øvrelid E, Bygstad B, Lie T and Bergquist M (2019) Developing and Organizing an Analytics Capability for Patient Flow in a General Hospital. *Information Systems Frontier* 22(2): 353–64.

Petrakaki D, Klecun E and Cornford T (2016) Changes in Healthcare Professional Work Afforded by Technology: The Introduction of a National Electronic Patient Record in an English Hospital. *Organization* 23(2): 206–26.

Raghupathi W and Raghupathi V (2014) Big Data Analytics in Healthcare: Promise and Potential. *Health Information Science and Systems* 2(1): 1–10.

Ranerup A and Henriksen HZ (2020) Digital Discretion: Unpacking Human and Technological Agency in Automated Decision Making in Sweden's Social Services. *Social Science Computer Review*. Epub ahead of print 17 December. https://doi-org.ezproxy.bib.hh.se/10.1177/0894439320980434

Roberts J (1991) The Possibilities of Accountability. *Accounting, Organizations and Society* 16(4): 355–68.

Rolandsson B (2019) The Emergence of Connected Discretion – Social Media and Discretionary Awareness in the Swedish Police. *Qualitative Research in Organisation and Management* 15(3): 370–87.

Rubel A, Castro C and Pham A (2019) Agency Laundering and Information Technologies. *Ethical Theory and Moral Practice* 22(4): 1017–41.

Sainsbury R (2001) Administrative Justice: Discretion and Procedure in Social Security Decision-making. In: Hawkins K (ed) *The Uses of Discretion*. Oxford: Oxford University Press, 295–329.

Shilo S, Rossman H and Segal E (2020) Axes of a Revolution: Challenges and Promises of Big Data in Healthcare. *Nature Medicine* 26(1): 29–38.

Susskind R and Susskind D (2015) *The Future of the Professions: How Technology Will Transform the Work of Human Experts.* Oxford: Oxford University Press.

Tresp V, Overhage JM, BundSchus M, Rabizadeh S, Fasching PA and Yu S (2016) Going Digital: A Survey on Digitalization and Large-Scale Data Analytics in Healthcare. *Proceedings of the IEEE* 104(11): 2180–206.

PART III

Experimenting with automation in society

10

HATE IT? AUTOMATE IT!

Thinking and doing robotic process automation and beyond

Martin Berg

Introduction

Picture a greyish office space with rows of desks cluttered with piles of paper and seemingly identical computer screens. Each desk is populated with a creature that brings back childhood memories: a red bobbing bird wearing a characteristic blue top hat. In this case, it does not mimic the motions of a drinking bird; instead, it moves up and down, rhythmically and almost as if in a state of trance, hitting its beak on the enter button of a keyboard. The scene comes from a commercial spot for the global automation software company UiPath (UiPath, 2020b). It places us in an unusually dull office space with appalling interior design. It invites us to put ourselves in the bobbing bird's position – a position where all things digital have gone awry. A narrating voice lets us know that 'digital transformation has failed to take off because it hasn't removed the endless mundane work we all hate'. Suddenly, one of the birds – presumably the one with which we are supposed to identify – stops picking the keyboard and turns its head toward a window where it stares at a real bird, with feathers instead of a top hat. The real bird is seemingly free from the dullness of office work. When the toy bird sees the other bird fly away, it puts on a confident face, as if it had listened to the narrator saying, 'automation can solve that by taking on repetitive tasks for us'. It jumps from the desk and takes off to a future that will help, again with the narrator's words, unleashing its potential.

This scene is one of many digital dramatisations of the future of work where all things 'hated' should be automated. A recently published report on tech trends for 2021 observes that many organisations across the globe are 'dragged down' by organisational debt caused by 'an extensive and expensive set of business processes underpinned by a patchwork of technologies that are often not optimised, lean, connected or consistent' (Burke, 2020). Across the globe, a vast array of companies tackle these challenges by offering systems and platforms for work automation,

DOI: 10.4324/9781003170884-14

mainly robotic process automation (RPA): a type of software that mimics human users when performing tasks in the graphical user interface of applications. According to recent reports, UiPath is the market-leading RPA platform, closely followed by competitors such as Blue Prism, Automation Anywhere, Workfusion and JIFFY.ai (Ray et al., 2020; SoftwareReviews, 2020). The need for products and services offered by companies in this growing field is often motivated with reference to a future where everything related to work is about to change due to the rapid advancement of automation technologies. This 'automation discourse', as Benanav (2019) labels it, involves dreams of human freedom, often connected to an idea of universal basic income to support what Bastani (2019) has labelled 'fully automated luxury communism', along with nightmares of mass unemployment as a consequence of fully automated 'lights-out' production and manufacturing (Frase, 2016; Ford, 2015; Srnicek and Williams, 2015). These kinds of ideas about the future of work, and indeed of humankind as such, should be understood as deeply embedded in modern capitalism. The production of such discourse needs to be approached critically as part and parcel of the socioeconomic systems in which automation technologies emerge and are presented as meaningful (Benanav, 2019).

Drawing on an analysis of the value propositions by two platform providers, UiPath and Blue Prism, this chapter unpacks the discursive practices that construct RPA as meaningful (Potter and Wetherell, 1987; Parker, 1992). UiPath (which is part-owned by Google Alphabet) and Blue Prism are two world-leading automation platform providers that offer companies across a wide array of sectors robotic support in different forms. Whereas UiPath offers 'a robot for every person' (UiPath, 2021g), Blue Prism presents its offer as 'a digital workforce' for the future (BluePrism, 2021e). Despite the minor differences in how these companies frame their offerings, they both employ a vocabulary that uses words such as change, re-imagination, reinvention, reboot and transformation. These accounts often take the form of evocative stories (Miller, 2007; Goode, 2018) and build on a mixture between technical specifications, future-oriented quotations and headlines, along with animated videos that illustrate work situations with and without RPA. The claims made are often *legitimised* by the continuous reference to the companies' white papers, webinars, case studies, endorsements and success stories from business partners across the globe.

Corporate actors within the field of work automation have adopted similar marketing strategies that revolve around storytelling on their web pages and social media platforms. They often share blog posts, instructional and promotional video clips, white papers, step-by-step guides and reports of different kinds. These materials provide a unique insight into how these companies present their fairly technical platforms to potential customers and how they imagine the future of work, with and without the presence of software robots and automated processes. The marketing materials that form the empirical foundation for this chapter are not only of a promotional or instructional kind. Instead, they rhetorically create value propositions by constructing a symbolic and somewhat imaginary context in which the promoted technologies seem to make perfect sense. That way, they produce a particular kind of knowledge about what work is and can become that is both situated and transformative. Engaging with discourse production of this kind is particularly important in critical studies of automation, as it can help us unpack built-in classification schemes, hierarchies and cultural assumptions that the system will treat as factual (Potter, 1996). This type of analysis can also help bring imaginaries of automation and human–machine collaborations into a more comprehensive discussion about what automated decision-making could mean and how it could be defined now and in the future.

Thinking and doing RPA

From the early discussions about computerisation and robotic technologies with cognitive capacities, automation technologies have often been discussed in terms of the possible futures of automation and how work tasks, as we know them, can be radically transformed: for good and for bad. The dreams, hopes and fears in response to automation can be traced back to the mid-19th century. They have influenced and inspired social theorists and scholars over the years, not the least as part of critical studies of capitalism. As Benanav (2019) observes, futuristic accounts of automation have emerged and re-emerged in the 1930s, 1950s, 1980s and again in the 2010s with utopian as well as dystopian promises and fears. Current technologies and platforms for work automation, mainly in the form of RPA and, to some extent, data-driven technologies fuelled by artificial intelligence (AI) and machine learning (ML), are discussed in ways similar to their historical predecessors, through an interplay between technocratic enthusiasm and critical pessimism (Bowler, 2017). These discussions have often focused on the relationship between automation, jobs and work tasks. Some imagine a future where traditional professions will vanish, while others argue that professions will only be partially automated (Brynjolfsson and McAfee, 2014; Frey and Osborne, 2017).

Researchers have debated the consequences of implementing digital workplace technologies since the 1980s (Grudin, 1994; Bowker et al., 2014). However, the implementation of work automation platforms such as RPA in already fairly complex techno-cultural workplaces is different from other computational support systems, since they are imagined to support people and fully or partially replace them (Manyika et al., 2017). Benanav (2019) distinguishes between automation technologies that 'fully substitute for human labour' and 'labour-augmenting technologies' that augment 'human-productive capacities' rather than replace workers and professionals within a specific job category. He points out that this distinction is a theoretical construct, since it is difficult to apply to real-world situations. As the discussion in this chapter shows, systems for work automation are often associated with multiple and somewhat contradictory expectations. They are believed to foster efficiency, productivity and precision, while at the same time allowing workers and professionals to spend less time on repetitive, rule-based and seemingly tedious work tasks. Workers are, therefore, supposed to invest more time and energy in their core professional practice and fulfil their personal life goals. These promises of automation technologies must be approached critically and understood as saturated with the biases and values of their makers: a group that often consists of a 'powerful elite of male white Silicon Valley engineers' (Wajcman, 2017: 123). These matters are often overlooked by the proponents of the automation discourse that, as Wajcman points out, avoid

> addressing the extent to which the pursuit of profit, rather than progress, shapes the development of digital technologies on an ongoing basis, and the ways in which these very same technologies are facilitating not less work but more worse jobs.
>
> *(2017: 124)*

These expectations do not emerge in a vacuum but rather result from personal imaginaries, experiences and expectations (Fors et al., 2016; Hornbæk and Hertzum, 2017; Fors et al., 2020). For that reason, 'the expectations, hopes, fears, and promises of new technologies are not set apart from, nor layered on top of scientific and technological practices, but are, rather, formative elements' (Selin, 2008: 1891; see also Urry, 2016). From such a perspective, it becomes clear that work automation platforms must be placed in a 'culture of anticipation' (Panchasi, 2009) rather than understood as a 'solutionist' answer to contemporary socioeconomic challenges (Morozov, 2013). As will be demonstrated in the following discussion, the value propositions of UiPath and Blue Prism are crafted with discursive practices through which their 'solutions' come to make sense. The futuristic and often promissory accounts of work automation rely on an understanding of work as flexible and somewhat malleable. The examples explored in this chapter tend to downplay that automation innovation and implementation are rather complex processes that transform work and work tasks in light of dominant discourses and structures. The following sections unpack some of the value propositions by UiPath and Blue Prism by exploring the envisioned raison d'être of work automation and notions of how they involve human activity.

Digital transformations gone awry

Visitors to UiPath's web pages are greeted by three cute and seemingly happy spherical robots equipped with antennas, bouncy legs and a head-mounted propeller that fit fairly well with their curious eyes looking invitingly at the visitor behind the screen. The robots are accompanied with a text saying 'Hello, we're UiPath. We make software robots, so people don't have to *be* robots' (UiPath, 2021a, original italics). The presence of such robots is supposed to 'help' the company 'show – in a simple, engaging way – how automation can do the work we humans hate, freeing us to do work that's more creative and rewarding' (UiPath, 2021i). The company gives the impression that it is indeed friendly, trustworthy and caring. It wants to take care of people at work by sharing 'knowledge' and offering a 'free global training resource' for RPA practitioners of different kinds, rather than simply selling products and services (UiPath, 2021k). UiPath claims to 'believe in creating a safe, generous, accepting workplace where people can be their authentic, best selves' (UiPath, 2021k) by allowing people to partake in a 'reboot' of work by using their automation platform. Referring to Bill Gates' vision of having 'a computer on every desk, and in every home', Daniel Dines, CEO at UiPath, envisions an automated future involving 'a robot for every person' (UiPath, 2021k).

Visitors to Blue Prism's web pages encounter a somewhat similar yet slightly more business-oriented visual approach. Under the heading 'Intelligent Automation', they show a looping video in which an animated female office professional stands still and looks into the camera (BluePrism, 2021c). She is smartly dressed in a yellow turtle-neck sweater and black trousers, wearing round yellow glasses that match the shape of her chic and corporate high bun hair. With one hand on her hip, and another on the thigh, she looks confident and professional, but something is missing from the picture: she does not smile. She does not even have a mouth – as if she were unable to speak. The imagery shifts, and the video shows the woman sitting in front of a computer screen, apparently experiencing the numerous apps and windows (email, spreadsheets and chats) as problematic and perhaps even chaotic. She points her finger up – as if she had an actual Eureka moment – and suddenly, the video shows how one of her documents becomes populated with pyramid-shaped robots that hover across the page, scanning and analysing every little detail of the document. The imagery switches again and presents us with the same woman, in the same pose, but this time surrounded by three robots in different colours. Suddenly, her mouth appears, and she puts on a smile to illustrate the accompanying text 'Unify your human and digital workforces. Free people to do great things' (BluePrism, 2021c).

These examples show how both UiPath and Blue Prism frame their automation platforms as invitations not only to transform work as such but also, and perhaps more importantly, to transform professionals and their lives. Mary Tetlow, Vice President of Global Brand Experience at UiPath, discusses these matters while commenting on the bobbing bird commercial spot that opened this chapter. In a backstory published on the company's blog, Tetlow writes:

> Over the years, I've spotted it sitting on office desks around the world. You probably have, too: the little toy bobbing bird. It dips down to take a sip of water, bobs upright, rocks back and forth, then dips down to take another sip. If you set it up right, it will do this again and again, repeating the same task. Day in, day out. It always made me smile, until the day I noticed that many of the people in those offices were doing the exact same thing. Stuck performing the same repetitive work tasks over and over. Day in, day out.
>
> *(Tetlow, 2020)*

Repetitive work tasks, Tetlow argues, result from a digital transformation that has given us more tools, systems and devices than we can use, thus adding complexity to our work lives rather than making them more manageable. By spending time fiddling with software and engaging in repetitive administrative tasks, people are believed to not 'feel productive, become dissatisfied, and lose motivation' (Tetlow, 2020). Systems for work automation are presented as solutions to such problems and are believed to provide people with an increased level of freedom by transforming work tasks' very nature. Tetlow continues to reflect on the possibilities with these systems, and she emphasises that '[p]eople are capable of so much more when they're empowered to do what humans do best: tackling the big problems'. Doing so, however, is thought of as almost impossible, since we must 'keep track of multiple pieces of technology at the same time' and 'nobody can multitask that well' (Tetlow, 2020). For this reason, UiPath claims that 'RPA is rewriting the story of work'. When software robots do 'repetitive and lower-value work' and 'high-volume tasks', people are 'freed to focus on the things they do best and enjoy more: innovating, collaborating, creating, and interacting with customers' (UiPath, 2021h).

Similar to Blue Prism's ambition to work toward 'intelligent automation', UiPath also offers robots that can engage in work tasks requiring 'cognitive processes' such as text interpretations and the application of 'advanced machine learning models to make complex decisions' (UiPath, 2021h). In order to make these proposals and ideas about work meaningful, the company shares an animated video labelled 'The story of work' in which we are told that work tasks – involving 'to grow things and create things and build things' – are part of human nature and 'what makes us happy' (UiPath, 2019). However, despite that 'people kept getting better at work' and that 'they built better tools to work more efficiently' things did not turn out as desired:

> They built amazing machines to work faster. They built computers to work smarter, but still, they couldn't do enough work because during all that work – to do more work, they had created work. And not the good kind! Not the let's sit around and come up with ideas kind. Not the "I can't wait to get to work so I can jump into a new project" kind. The crap kind! The drudge and data and admin and the damned expense reports. . . . So the humans had to work more at work they didn't like. That made them robotic. That didn't inspire. They had to postpone vacations, they had to miss family dinners, they were pissed!
>
> *(UiPath, 2019)*

The introduction of RPA should 'put the fun back into work' and make people 'waste less time doing the things they hated' so that humans could stop acting like robots and instead start teaching robots how to make people free – 'Work is history. It's time to reboot work' (UiPath, 2019). A similar line of reasoning is present in one of Blue Prism's promotional videos, suggesting that work tasks should match human nature. Under the heading 'Get Work Done with Intelligent Automation', they explain this idea further:

> Today, work means doing many different tasks: reading emails, creating spreadsheets, and even using that irreplaceable twenty year old program. It's a lot of work, and not all of it uses the skills that make us, well, human! Like creativity, critical thinking, and communication. So why not automate those repetitive tasks and focus instead on the high-value great work we are meant to do? With the aid of Blue Prism's 'digital workforce' people are supposed to be allowed to

focus on work tasks that are meaningful and valuable while using their human capacities to focus on what truly matters for them. This involves, however, that new forms of collaboration emerge, and indeed also new work tasks.

(BluePrism, 2021b)

These accounts build on an idea that something is fundamentally wrong with contemporary work and workplaces due to a digital transformation that has not lived up to what it promised. Instead of discussing efficiency, profit and potential structural and economic factors behind this state of affairs (in short: capitalism), UiPath and Blue Prism motivate the need for work automation platforms in affective and sociocultural terms. The return on investment in these cases is often described by referring to people becoming happy, content and personally fulfilled by not engaging in specific work tasks: the boring, repetitive and pointless ones. However, rather than simply allowing workers and professionals to do what they believed to do best, implementation of work automation involves a professional transformation through which new and higher valued work tasks should be learned. Automation platforms thus involve more than simply adding 'a digital workforce' or a 'robot for every person'. They also involve and require an organisational restructuring through which a digital transformation that fits better with the imaginaries of our times is done.

Working with unleashed potentials

'What if there was an innovative technology that truly improved productivity and let people focus on things they really enjoy doing or are great at?' UiPath poses this question in one of their promotional videos as a response to the observation that people spend a fair amount of time using technologies and software to deal with 'mundane tasks' and 'work that doesn't bring any value' (UiPath, 2020a). As an alternative to spending hours on such work tasks, the company invites their potential customers to

> [i]magine how much more productive it would be if every employee at your company had their own robot assistant to do the busy work so they could work faster on higher value tasks that make them happier and maximize the impact on your business!
>
> *(UiPath, 2020a)*

As they label the idea, this 'innovative vision for the future of work' is key to understanding how UiPath frames its software offerings. UiPath claims that its RPA platform allows 'companies to automate routine tasks using software robots that emulate humans, so employees spend less time on manual work and more time on activities that leverage their valuable – and uniquely human – skills' (UiPath, 2021b). UiPath describes RPA as a software technology used to create and manage 'software robots that emulate human actions interacting with digital systems and software' (UiPath, 2021k). However, they are believed to do so in a way that goes beyond the imagined limitations of people:

> Just like people, software robots can do things like understand what's on a screen, complete the right keystrokes, navigate systems, identify and extract data, and perform a wide range of defined actions. But software robots can do it faster and more consistently than people, without the need to get up and stretch or take a coffee break.
>
> *(UiPath, 2021k)*

The company claims that its purpose is to '[a]ccelerate human achievement' since they 'see boundless potential in the way we live' and 'believe in using the transformative power of automation to liberate the boundless potential of people' (UiPath, 2021k). Blue Prism presents a similar idea in a futuristic promotional video that puts us right in the middle of an urban digital environment that creates a feeling of what an automated future might hold. 'Welcome to the art of the possible!' says the video's narrator, in which the company presents its 'Intelligent Automation' vision for a 'world propelled by powerful automation technologies'.

Similar to the previous example in this chapter, Blue Prism offers a platform for RPA that becomes 'intelligent' through integration with technologies such as AI and ML that go beyond ruled-based processes and thinking. At the core of this platform lies what they label 'digital workers' – 'super organized, multitasking software robots that work alongside your people to automate and transform business process' (BluePrism, 2021d). In the video, animated vector illustrations of robotic bodies with colourful highlighting of their brains, interconnected and hard-working in front of computer screens represent the digital workforce. The narrator invites potential customers to compare the digital workforce with 'traditional' workers:

> Like humans, digital workers can develop new skills over time, getting smarter and more capable. With AI, Blue Prism digital workers can be trained to take on increasingly complex tasks, manage vast workloads, and make critical decisions to tackle work with greater speed and productivity, becoming a force multiplier in your business.
>
> *(BluePrism, 2021d)*

With Blue Prism's 'digital workforce' assistance, people should be able to avoid engaging in 'Time-consuming tasks that don't necessarily create value. And those tasks don't all use the skills that make us human' (BluePrism, 2021b). The addition of digital workers does not mean that the automated processes should be autonomous, but rather that 'traditional' and 'digital' workers should 'work side-by-side, giving /. . ./ people more time to focus on strategic, meaningful work' (Blue-Prism, 2021a). This automation platform involves three phases for working with the system: discover, design and deliver. By identifying which business processes to automate and then building the actual workflows, potential users are told that the first steps of implementing the Blue Prism platform 'can seem like a plug-and-play operation'. The instructions seem pretty straightforward: 'Just think about – and articulate – the steps your people are performing today, and you're set'. In order to move further by 'building a sustainable and scalable digital workforce', however, one must know that such an operation 'takes a long-term delivery strategy and the right tools and technology to make it a reality' (BluePrism, 2021a).

Similarly, the UiPath platform consists of several products that all, in different ways, serve not only at automating enterprises but indeed also to digitally transform them. The software platform consists of five parts that each represents different work tasks related to their software robots. Potential users are invited to use the platform to discover, build, manage, run and engage. Although the software robots are presented as means to take away work tasks, to free up time and to allow people to engage in tasks that resonate with their uniquely human qualities, as described in this chapter's introduction, they do so, interestingly enough, by inviting potential users to new and different work tasks. For instance, the UiPath platform allows users to engage in 'automation discovery' through which they are supposed to learn about their business processes and understand how people in the enterprise work. Such a discovery involves gathering 'automation ideas' from employees across the organisation or visualising automation processes and tasks.

Similar ideas apply to the other areas of UiPath's platform. Employees are invited to not only use robots but indeed also to build them. As the company announces on their web page, 'Everyone can be an automation creator or contributor. Think of the productivity!' (UiPath, 2021j). In addition to creating instances of automation – a practice where employees become, as UiPath has chosen to label it, 'citizen developers' – people across the organisation are allowed to build, engage with and manage robots in different ways. Put differently, they can '[b]uild apps in a snap, deploy with a click, with no coding' (UiPath, 2021j). In UiPath's promotional materials, potential users are constantly invited to see new and different work tasks as opportunities and alternatives to their current work, rather than being liberated from work as such.

It should be clear that implementing an automation platform requires working with that same platform in ways that involve interaction, maintenance and even repair (Jackson, 2014; Puig de la Bellacasa, 2017). The two platforms explored in this chapter include 'intelligent' components and add-ons that are supposed to assist people in handling and working with the automation processes. However, the level of automation or the extent to which artificial intelligence or machine learning drives these processes is relatively unimportant for understanding these platforms and their promises from a critical perspective. More important is how *doing* automation – rather than simply relying on its support – is a question of a far-reaching transformational process or a 'journey', as UiPath labels it. According to UiPath, a successful 'RPA journey' should follow a '3-stage path' to 'make sure you wind up in a great place'. Such a transformational 'journey' is not primarily about writing robotic scripts to manage work tasks. Instead, it involves putting 'more joy in each employee's workday by taking repetitive routine [sic] out of it' through a fundamental organisational transformation toward becoming 'more efficient, agile, and profitable' (UiPath, 2021d).

The trajectory that UiPath has set out for companies to become successful in their RPA implementation starts with proof of concepts 'to get the buzz started' and to 'test the waters' (UiPath, 2021f). While the first stage aims at finding the 'ripest pipeline opportunities: easy-to-implement automations with high ROI', the second step moves toward the development of 'complex cross-enterprise processes' that involve 'core systems, important functional areas, and key enterprise activities' (UiPath, 2021c). The third and the final step of the automation journey is where the 'dream of "a robot for every person" [will] come true'. This step consists of training 'people on the technology', 'encourage innovation and ideas for projects' and launching 'a citizen developer program that provides technology and training to everyday business users so they can reduce repetitive drudgery in their everyday tasks' (UiPath, 2021e). This final stage is mainly about making people and robots come together by spreading 'robot love' and by helping 'everyone fall in love with their new digital assistant through training and change management' by tapping 'into people's desire to learn how to automate' (UiPath, 2021d). These examples show that implementing a platform for work automation does not only build on and indeed require a particular taxonomy of work tasks. It also requires that the very nature of work – no matter the kind of business or organisation involved – needs to be reimagined and re-constructed.

Re-constructing automated futures

Suppose automation is the solution to an identified problem. One must ask what that very problem consists of, under what circumstances it exists and what happens once it is solved. The previous sections explained that RPA is more than a technology or system for process automation; it is also a sociotechnical prism through which the past, present and future of work are imagined and re-constructed. Most, if not all, readers of this book will likely share the experience of work as not always being particularly meaningful or creative to the extent that it brings out our inherently human qualities. Work tasks come in many forms and flavours, some of which might be considered dull, repetitive and potentially pointless, and others that could be understood as creative and meaningful for some but alien for others. Experiences of work and work tasks depend on perspective and context, but the accounts of RPA explored in this chapter often overlook this state of affairs. Promissory technologies such as RPA thus require discursive practices that create a particular taxonomy of work tasks. Some work tasks are constructed as meaningful and desirable, and others as a pointless reminiscence of previous digital transformations that have gone sour. The claim that some work tasks do not resonate with human nature functions as a lure for the proposal for a fundamental organisational transformation through which all work tasks should be examined and re-evaluated against how an automated future might look. As the earlier discussion has shown, systems for work automation are often defined by constructing a future scenario that requires that the present is redefined or reimagined as problematic in a particular way.

The empirical examples in this chapter show that RPA requires a particular way of understanding work tasks. Some tasks must be regarded as boring (mostly the repetitive ones) and others (mostly the ones involving creativity) need to be positioned as meaningful. This separation between different work tasks is intriguing, since the RPA offerings involve more than the technologies and systems. They also involve practices through which the domains where these technologies can be implemented are described, defined and – to some extent – invented. Not only do these domains need to be created, but they must also be constructed as problematic, dysfunctional and in need of transformational repair. Such a transformational process – or a rebooting, to use UiPath's words – follows a trajectory marked by particular ideas about creativity and what counts as meaningful work. It follows pretty naturally that a specific understanding of the present is given form by constructing a particular future, but importantly these accounts evade questions of capitalism as the socioeconomic system in which the automation discourse is deeply embedded (see also Spencer, 2018).

As described in this chapter, work automation, regardless of whether it takes place in private or public sectors, involves transforming work tasks inspired by the creative industries. Such a transformation regards design thinking, agile methods, creative workshops and tinkering with scripts and robots as high-level tasks meaningful for people, despite professional roles and experience. However, asking how these ideas play out across different sectors and businesses is appropriate. What happens when the public sector in a particular country with its legislation and administrative tradition, to take an example, becomes structured following principles and ideas from tech companies and the creative industries? When work automation is carried out as a practice rather than implemented as a supportive technology, work tasks are required to gravitate around an axis of design thinking that should be more familiar to people in the creative industries than in, for instance, the public sector. The fact that work automation involves more than simply getting a nifty tool is essential for how interaction and relationships between people and machines – or, as Blue Prism chooses to put it, between traditional and digital workers – are envisioned. Throughout the examples mentioned earlier, the robots have been considered as 'buddies' and friendly little helpers with which people are supposed to establish affective bonds: almost as if they were friends or close colleagues that both assist and become guided by 'traditional' workers.

Concluding remarks

At the outset of this chapter, a commercial spot from UiPath was presented. The video, revolving around an office-bound bobbing bird that decided to break free from the dullness of repetitive administrative tasks, was most likely supposed to bring back childhood memories and act as a playful lure to make people desire something different from their regular work life. Having explored the value propositions by UiPath and Blue Prism, it has become clear that the bobbing bird, by taking off into an automated future, would most likely encounter more than a feeling of being set free. It would also have to rethink work and the potentials of automation. Once it had landed in a new and different work reality, it would also have to start engaging in work tasks that would be different from pecking at the enter button. However, there is no guarantee that the creative and more 'human' work tasks (which might result as challenging even for a bobbing bird) are less repetitive and mundane than those from which it fled. Even fiddling with robot scripts, mining tasks for automation and sharing automation proof of concepts among colleagues might get boring and monotonous in the long run. Perhaps because, as Spencer (2018: 1) suggests, 'the quest for a more humane work environment – one that supports extended free time while encouraging more intrinsically rewarding work – requires changes in ownership that cede power to workers over the use of technology'. No matter the validity of this suggestion, the examples in this chapter show that the study of automation technologies and their promissory character require a robust research agenda to unmask critical assumptions and potential power relations involved in how the development and implementation of such technologies are envisaged and performed. Such an agenda would consider the socio-technological frameworks or contexts for such technologies and how discursive practices produce those frameworks and contexts.

Acknowledgement

The chapter derives from the research project 'Working with Algorithmic Colleagues: Expectations and Experiences of Automated Decision-Making', funded by the Swedish Research Council (grant number 2020–00977).

References

Bastani A (2019) *Fully Automated Luxury Communism: A Manifesto*. London and New York: Verso Books.

Benanav A (2019) Automation and the Future of Work – I. *New Left Review* 119: 5–38.

BluePrism (2021a) *Blue Prism Product Portfolio*. Available at: www.blueprism.com/products/ (accessed 11 July 2021).

BluePrism (2021b) *Get Work Done With Intelligent Automation*. Available at: www.blueprism.com/resources/videos/get-work-done-with-intelligent-automation/ (accessed 11 July 2021).

BluePrism (2021c) *Intelligent Robotic Process Automation – RPA*. Available at: www.blueprism.com (accessed 11 July 2021).

BluePrism (2021d) *Six Intelligent Automation Skills*. Available at: https://video.blueprism.com/watch/LPQEFcLMNaNB3DcGFgBKyF? (accessed 11 July 2021).

BluePrism (2021e) *Way beyond RPA: Intelligent Automation for the Enterprise*. Available at: www.blueprism.com/uploads/resources/Blue-Prism-Corporate-brochure-2021.pdf (accessed 1 July 2021).

Bowker GC, Star SL, Turner W, et al. (2014) *Social Science, Technical Systems, and Cooperative Work – Beyond the Great Divide*. New York: Psychology Press.

Bowler PJ (2017) *A History of the Future: Prophets of Progress From H. G. Wells to Isaac Asimov*. Cambridge: Cambridge University Press.

Brynjolfsson E and McAfee A (2014) *The Second Machine Age: Work, Progress, and Prosperity in a Time of Brilliant Technologies*. New York and London: W. W. Norton & Company.

Burke B (2020) Gartner Top Strategic Tech Trends for 2021. Available at: www.gartner.com/en/information-technology/trends/2021-top-strategic-technology-trends-gb-pd-brnd.html (accessed 22 March 2021).

Ford M (2015) *Rise of the Robots: Technology and the Threat of a Jobless Future.* New York: Basic Books.

Fors V, Berg M and Pink S (2016) Capturing the Ordinary: Imagining the User in Designing Automatic Photographic Lifelogging Technologies. In: Selke S (ed) *Lifelogging: Digital Self-tracking and Lifelogging – Between Disruptive Technology and Cultural Transformation.* Wiesbaden: Springer VS, 111–28.

Fors V, Pink S, Berg M and O'Dell T (2020) *Imagining Personal Data: Experiences of Self-tracking.* London and New York: Routledge.

Frase P (2016) *Four Futures: Life After Capitalism.* London and New York: Verso Books.

Frey CB and Osborne MA (2017) The Future of Employment: How Susceptible are Jobs to Computerisation? *Technological Forecasting and Social Change* 114: 254–80.

Goode L (2018) Life, But Not As We Know It: A.I. and the Popular Imagination. *Culture Unbound* 10(2): 185–207.

Grudin J (1994) Computer-supported Cooperative Work: History and Focus. *Computer* 27(5): 19–26.

Hornbæk K and Hertzum M (2017) Technology Acceptance and User Experience. *ACM Transactions on Computer-Human Interaction* 24(5): 1–30.

Jackson SJ (2014) Rethinking Repair. In: Gillespie T, Boczkowski P and Foot K (eds) *Media Technologies: Essays on Communication, Materiality, and Society.* Cambridge: The MIT Press.

Manyika J, Chui M, Miremadi M, et al. (2017) A Future That Works: Automation, Employment, and Productivity. McKinsey Global Institute, San Francisco. Available at: http://www.mckinsey.com/global-themes/digital-disruption/harnessing-automation-for-a-future-that-works (accessed 17 March 2022).

Miller R (2007) Futures Literacy: A Hybrid Strategic Scenario Method. *Futures* 39(4): 341–62.

Morozov E (2013) *To Save Everything, Click Here: Technology, Solutionism, and the Urge to Fix Problems That Don't Exist.* London: Penguin Books.

Panchasi R (2009) *Future Tense: The Culture of Anticipation in France Between the Wars.* Ithaca, NY: Cornell University Press.

Parker I (1992) *Discourse Dynamics: Critical Analysis for Social and Individual Psychology.* London and New York: Routledge.

Potter J (1996) *Representing Reality: Discourse, Rhetoric and Social Construction.* London: SAGE.

Potter J and Wetherell M (1987) *Discourse and Social Psychology: Beyond Attitudes and Behaviour.* London: SAGE.

Puig de la Bellacasa M (2017) *Matters of Care: Speculative Ethics in More Than Human Worlds.* Minneapolis, MN: University of Minnesota Press.

Ray S, Villa A, Tornbohm C et al. (2020) Gartner Magic Quadrant for Robotic Process Automation. Available at: www.gartner.com/en/documents/3988021 (accessed 22 March 2021).

Selin C (2008) The Sociology of the Future: Tracing Stories of Technology and Time. *Sociology Compass* 2(6): 1878–95.

SoftwareReviews (2020) Robotic Process Automation Data Quadrant Report. Available at: www.softwarereviews.com/reports/Category/206/category/categories/async_offerings_load/91575/download (accessed 22 March 2021).

Spencer DA (2018) Fear and Hope in an Age of Mass Automation: Debating the Future of Work. *New Technology, Work and Employment* 33(1): 1–12.

Srnicek N and Williams A (2015) *Inventing the Future: Postcapitalism and a World without Work.* London and New York: Verso Books.

Tetlow M (2020) What a Simple Office Toy Taught Me About Humans. Available at: www. uipath.com/blog/digital-transformation/what-simple-office-toy-taught-me-about-humans (accessed 11 July 2021).

UiPath (2019) The Story of Work. Available at: https://youtu.be/dJ9c2xq6mas (accessed 11 July 2021).

UiPath (2020a) A Robot for Every Person: Benefit from #Automation at Enterprise-Wide Scale. Available at: https://youtu.be/D7A2J054wxM (accessed 11 July 2021).

UiPath (2020b) Take Off. Available at: https://youtu.be/eIpdyTqjjQ0 (accessed 11 July 2021).

UiPath (2021a) Automation Platform. Available at: www.uipath.com (accessed 11 July 2021).

UiPath (2021b) Charting the Automation Journey: A 3-Stage Model. Available at: www.uipath.com/solutions/whitepapers/charting-automation-journey (accessed 11 July 2021).

UiPath (2021c) Go Wider, Higher, Deeper by Automating Key Cross-Enterprise Processes. Available at: www.uipath.com/rpa/journey/scaling-automation-across-enterprise (accessed 11 July 2021).

UiPath (2021d) How to Start Your RPA Journey . . . and Make Sure You Wind Up in a Great Place. Available at: www.uipath.com/rpa/journey (accessed 11 July 2021).

UiPath (2021e) Make the Dream of 'A Robot for Every Person' Come True. Available at: www.uipath.com/rpa/journey/providing-software-robot-for-every-person (accessed 11 July 2021).

UiPath (2021f) Prove Out, Build Out, Move Out. Available at: www.uipath.com/rpa/journey/getting-started-automation-journey (accessed 11 July 2021).

UiPath (2021g) A Robot for Every Person: Empower People to Work with Software Robots. Available at: www.uipath.com/rpa/robot-every-person (accessed 11 July 2021).

UiPath (2021h) Robotic Process Automation (RPA). Available at: www.uipath.com/rpa/robotic-process-automation (accessed 11 July 2021).

UiPath (2021i) Style Guidelines|2.6. Available at: www.uipath.com/hubfs/Valentin/Brand-Kit/UiPath-brand-guidelines.pdf (accessed 11 July 2021).

UiPath (2021j) The UiPath Platform: Become a Fully Automated Enterprise™ with the UiPath Platform. Available at: www.uipath.com/product (accessed 11 July 2021).

UiPath (2021k) The UiPath Purpose: Accelerate Human Achievement. Available at: www. uipath.com/company/about-us (accessed 11 July 2021).

Urry J (2016) *What Is the Future?* Cambridge: Polity Press.

Wajcman J (2017) Automation: Is It Really Different This Time? *The British Journal of Sociology* 68(1): 119–27.

11
SMART THERMOSTATS AND THE ALGORITHMIC CONTROL OF THERMAL COMFORT

Julia Velkova, Dick Magnusson and Harald Rohracher

Smart temperatures

Smart speakers, doorbells with facial recognition, personal smart entertainment systems, digital home assistants, smart lighting and many more devices have all become part of a growing landscape of media that add a layer of 'local intelligence' (Thrift and French, 2002) to domestic space and infrastructure. Smart thermostats are a further example of such devices. They are gaining popularity in everyday life through Big Tech's 'smart home' systems such as Google Nest and Apple's HomeKit, and are available in many versions, sold by several companies. They promise to provide users with increased temperature control and thermal comfort at home, while at the same time-saving energy. The Google Nest Learning Thermostat, for example, is presented as a device that 'learns the temperatures that you like when you're at home and then programs itself. It automatically turns down the heating when you're away to help save energy' (Google Nest Help, n.d.). Such marketing promises activate an old techno-utopian ideal of providing users with digital servants who can care for their needs and desires in optimal, individualised and pre-emptive ways (Suchman, 2007: 219). In practice, they remake the spaces of everyday life into programmable environments and try to configure domestic life and infrastructure to perform optimally through information and sensing (Gabrys, 2014).

As consumers buy smart thermostats and try to make their homes more 'intelligent', energy utilities in many countries have discovered the potential of 'moving beyond the metre' and are redefining their relations with energy customers. 'Smart home' technologies – with smart thermostats as part of this ecology of devices – have made it possible for utilities to offer a broader range of services (e.g. real-time energy-consumption feedback) and to harness the flexibility that households have to shift some of their electricity or heat use during the day to optimise their own energy system. Despite such attempts to steer energy use patterns in homes, users

DOI: 10.4324/9781003170884-15

are promised that their freedom to control their thermal home environment will not be compromised. Against this backdrop, some public and private energy companies have started to experiment with adding automated decision-making (ADM) to smart thermostats to both track users' thermal preferences and everyday energy-use routines, take decisions about controlling their thermal comfort in real time and ensuring cost-efficient operation of the local energy grid.

In this chapter, we discuss how energy companies and tech start-ups are experimenting with embedding ADM systems into smart thermostats, simultaneously testing and reshaping relations between private homes, energy infrastructure providers and data-driven companies. Our approach is inspired by the recent call (Marres and Stark, 2020) within science and technology studies (STS) to re-invigorate the experiment as an empirical and theoretical lens that can help us to understand how testing operates on and reshapes social life. As Marres and Stark (2020: 433) argue, experiments in digitally mediated contexts are not done *in* social environments but rather 'the social environment is itself the object of testing'. Drawing on this perspective, we suggest that ADM should be seen not as a process of delegating decisions from humans to machines but as a mediating algorithmic logic that, through experiments, binds together, mediates and transforms relations between multiple economies and 'social worlds' brought together by a common concern with temperatures. We develop this argument in the empirical context of an experimental project, Thermo-S, that took place in Sweden in 2019–2020. The next section presents the experiment and our approach in more detail.

Experimenting with ADM

Thermo-S was the name of an experiment that took place in a popular Swedish mountain resort, Åre, between 2018 and 2021. Our knowledge of it came through media publicity, where it was described as 'a unique self-learning project' that was 'first in the world to digitalise district heating' (Jämtkraft, 2018) and introduce ADM as a service to energy companies and homeowners. The experiment was realised as a partnership between a local, publicly owned district heating (DH) company, Jämtkraft, and a Swedish producer of smart thermostats, Ngenic. Participants in the experiment were provided with smart thermostats and promised advanced thermal control of their homes, while at the same time an ADM system at the backend of the thermostat would incrementally steer home temperature according to the needs of the energy company to optimise its DH infrastructure. Temperatures were controlled by algorithms that used data and sensors installed in participants' homes to demonstrate the efficiency of automatic steering. The smart thermostats are commercially available, but within Thermo-S they were provided free of charge to 21 volunteers and were installed in 37 buildings, including a few hotels, apartments and houses. It was intended that the experiment, if successful, would pave the way for a large-scale roll out of these devices and algorithmic steering of temperatures in other DH systems owned by the energy company.

Thermo-S reflects contemporary visions about the creation of 'smart', programmable environments, infrastructures and homes, taking shape today through sociotechnical experiments in 'real-world' settings. As scholars within STS have argued, such experiments operate at multiple scales (Ansell and Bartenberger, 2016) and are public demonstrations of the ideas. They are intended to persuade selected audiences, even when their outcomes are largely uncertain (Laurent, 2016). In such experiments, computational means are used to modify society (Marres, 2020) and to turn social life itself into an object of observation, quantification and testing by a wide variety of actors (Engels et al., 2019; Marres and Stark, 2020). Marres and Stark (2020) have more recently suggested that the long-standing concern of STS with sociotechnical experiments should be extended to conceptualise such experiments not as settings within which something is to be confirmed, proved or established as a 'fact' but as generative environments through which new relations between actors are established, alongside new modalities of knowing, valuing and acting.

Drawing on this perspective, we regard ADM in the context of the Thermo-S experiment as generative of a new sociotechnical environment in which new relations between households, the energy company and the thermostat provider are being 'figured out' and constituted. We were not part of Thermo-S but studied it at the end of the experiment with the aim to understand how it had unfolded in everyday life, what experiences, conflicts and frictions emerged in relation to the ADM steering, and what broader implications of the experiment arose. We analyse in this chapter how the experiment was understood in the distinct 'social worlds', economies of value and social practices to which the actors gathered in the experiment belonged, that is home dwellers, an energy provider and a start-up company producing IoT solutions and data services. While the aim of the experiment was to determine whether an ADM system for algorithmic temperature steering could bind together all three social worlds in a way that was 'optimal' to each of them, we analyse here how it functioned and how it was valued differently by each of the actors.

Methodologically, we draw on a combination of interviews and digital methods. We interviewed two engineers in charge of the experiment, one from Jämtkraft and one from Ngenic, in early 2021. We approached all 21 building owners and interviewed six of them in the spring of 2021, to obtain their perspectives. Two of them owned multi-tenant buildings that are usually rented out to tourists, and the rest were single-family homeowners. In addition, we performed a version of the walkthrough method (Light et al., 2018) on the app that the participants received and consulted marketing and press materials available online.

We discuss in the following sections the three realms and their different modes of valuing and relating to the experiment separately and analyse the frictions and new relations that arose.

Thermal comfort in the home

The setting for testing the ADM algorithm in the Thermo-S experiment was the home and the thermal comfort it provides to its dwellers. Comfort and energy use in homes have attracted the attention of social science research for a long time (see, e.g. Shove, 2003; Gram-Hanssen, 2010). In this context, indoor temperatures have always implicitly raised questions of social control and transformation (Beregow, 2019). The politics of control and manipulation of thermal systems have historically been associated with the engineering of technical solutions that provide pleasure and entertainment in everyday life and a sense of modernity and progress (Ackermann, 2010). As Cooper (2002) has shown, the installation of air conditioning in American homes has also led to the redesign of houses and changes in the standard of buildings and in cultural expectations of everyday life in artificially temperate environments. These expectations were shaped by engineering visions and practices while users were generally assigned a passive role in these developments. This changed dramatically with the advent of devices marketed through the discourse of 'smart homes', which emphasised the 'activeness' of users and their capacity to automatically control the home and its indoor environment – a user imaginary that was succinctly labelled 'resource man' by Strengers (2013).

Imaginaries of smartness and temperature control also played an important role in the Thermo-S experiment. In order for experiments to function, social settings must be 'instrumented' and curated. Such curation requires mediation, a materialising practice through which devices, apps, buttons and links are inserted into social environments, with the aim of rendering them representable and liable for actionable, establishing connections between a test 'field' and a 'laboratory' (Marres and Stark, 2020). As part of instrumenting the ADM intervention, Thermo-S required that the participants install smart thermostats in their homes, as well as an app developed by Ngenic. The participants were asked to activate and use these devices through which their thermal comfort would be algorithmically steered.

The Ngenic smart thermostat was presented to participants as working to improve the thermal comfort in the home. The system's built-in 'intelligence' would track and predict individual comfort needs; it would 'learn' temperature preferences at different times of the day, keep participants informed about the current state of their indoor environment, and give them control, also remotely. The latter function allowed them to turn on or off the heating at home while sitting in the train or car on their way home. Marketing materials suggested that the devices were 'extra brains for heating your house', 'self-learning systems that help you save energy and money – so that you can do something else'. Ngenic promised to make users into 'climate heroes', while ensuring an individualised and increased sense of comfort, and a consistent indoor temperature (Ngenic, n.d.).

In the Thermo-S experiment, the degree of agency of participants over their thermal comfort was negotiated through the app that came with the smart thermostats. Participants must first activate the app, after which they were allowed to express their desires for a comfortable thermal environment at home. While installing the app, the participants consented to 'terms of service', through which they delegated the control of their heating system to Ngenic's ADM system. These terms of service also assigned responsibilities – to Ngenic to algorithmically steer and track energy use in the home in a way that it perceived as efficient, and to participants to support the system by ensuring that the smart thermostats were always connected to the internet. Crucially, the terms of service suggested that the efficiency of steering was based not only on users' preferences but also on 'external demands':

> Ngenic has the right to steer based on external demands, for example but not limited to grid capacity reserves, balancing power in order to enable an efficient energy use and energy system. Such steering is limited so that the users' comfort is not influenced in a meaningful way. Ngenic can occasionally extend the steering and should inform then the user of eventual discomfort.
>
> *(Ngenic – Terms of Service, n.d.)*

While the language of intelligence sought to create a sense of empowerment of the users and suggest a sense of control, what participants in Thermo-S were actually doing was to invite a third party to track their thermal preferences and control them according to its own ideas of efficiency.

Unsurprisingly, the user interface of the app did not allow for much control – the main thing that users were encouraged to set was the temperature between 12 and 24 degrees, using the app as a remote control. They were also shown simple charts that presented statistics over the change of daily indoor and outdoor temperature. To make this rather limited form of user control seem more dynamic and compelling, the app applauded participants when they changed the desired temperature by 0.5 degrees or more, with acclamations such as 'Woop, Woop! Your changes have been saved!', assuring them that the system nevertheless works *for them*. However, participants were rather passive in practice, and few changed any settings. One said: 'I never change or touch anything, I have set it up where I want to have the device in the house. I have set it at 21, or 20.5 degrees. And it works'. Another participant reflected more on how the algorithmic control behind the app worked, but overall was also rather passive:

> Well, I never change anything. The thermostats stay where they are, and the app stays too. But what I wonder about sometimes is whether it matters at all what I do with the thermostats at home. It shouldn't matter whether we control the thermostats or not, but I feel that it does matter.

Some participants were only checking the temperatures displayed on the app screen. One explained: 'I use the app mostly to look at the temperature outside, as I know that it is correct'. Others did not find much meaning in looking at the temperature statistics at all: 'No, no, I haven't followed any charts. I see that there are charts. But I don't sit and watch them'. Others appreciated the possibility to set the temperature of their houses remotely:

> It is nice that while we live in Stockholm, we can open the app every now and then and check the temperature, that it stays where it should be. We plan then to travel up to Åre on the 7th or 8th. And I have set the app so that it

should be warm in the house a day before. And then you can actually see that it starts [heating up] even maybe a day before that.

Using the app as a remote control gave some users the feeling that they were in control of their thermal environment, and that they could reduce their heating costs. But, as Benson-Allott (2015) points out, while remote controls are the most common media device of the 21st century, they are both a technology and a cultural fantasy. They give a 'push-button sovereignty' – a sense of control over our personal media technologies that builds upon historically specific ideas about how users should interact with the industries 'behind' the remote, while at the same time limiting sovereignty over these devices. In the Thermo-S experiment, participants could scroll and select temperatures remotely as much as they wished, but the DH company and the new ADM algorithms experimented and controlled their heating according to their own rationalities and needs. Although the smart thermostats seemed to shift control and action to participants, independence in temperature control remained a cultural fantasy, while the ADM system also let in others to occupy the virtual control room of the home, as discussed later.

Optimising infrastructures

While from the perspective of the participants, the smart thermostat from Ngenic gives advanced control over the thermal environment at home, this picture looks very different when seen through the eyes of the energy company that is in charge of providing DH to houses. Its concern is the limited capacity of the DH grid and the need to cover peaks in demand with heat from fossil fuels. While the mountain resort has grown, and cultural expectations for heat delivery at particular times have evolved, the energy company has been compelled to become innovative in order to keep up with demand while at the same time generating profit.

Demand-side management by the energy company can be achieved in several ways. In the electricity grid, time-dependent tariffs can be offered to users, making energy more expensive at certain times of the day. Peak periods are, for example, in the morning when many use warm water for showering, and in the afternoon when saunas are turned on for 'after-ski'. Another approach is to allow the energy company to switch off certain devices, such as heat pumps, for brief periods when demand is high. Experimenting with ADM via the smart thermostats in Thermo-S allowed the energy company to track and control users, to borrow and redistribute heat between their homes via the DH system and to predict heat demands from earlier data without changing the physical infrastructure. In advance of an expected peak (such as the period of early morning showers), the room temperature in the homes with the Ngenic thermostat could be raised slightly, so that the temperature in bathrooms and other rooms could subsequently be lowered during the peak period, reducing the overall need for the utility to produce heat.

By avoiding such fluctuations, the DH system is shaped to operate under an economic rationality of 'optimal' infrastructure use, while the ADM algorithms also carry the promise for the energy company to avoid investing in new production plants and larger pipes, reducing the need to burn oil. As an engineer from Jämtkraft explained:

> We want to keep an eye on our client from a production perspective and from the perspective of the user's heat usage patterns, in order to predict how we should run our heat furnaces in order to make an optimal and stable network.

Keeping an eye on citizens also meant algorithmic control of their heating and their experience of thermal comfort. This rationality in the Thermo-S experiment required that 'users' be redefined. For a long time, households connected to DH in Sweden have been conceived as 'offloading heat points' by the utilities, without the possibility to significantly influence heat use in homes: 'Our infrastructure was too rigid to react to or guide user behaviour', a Jämtkraft engineer admitted. Ngenic's smart thermostats allowed the DH utility to redefine users into 'active nodes', whose thermal comfort could be steered towards particular aims of the utility: 'Steering the nodes produces much higher cost efficiency, because you can do much more with your existing infrastructure, bypassing local and global limitations in the system', one Ngenic engineer explained, on behalf of the energy company Jämtkraft. Still, this new system of control required the acceptance and cooperation of users. Ngenic defined this cooperation to the utility in the following way:

> And, here we should remember that in order for the client to consider this meaningful, they need to get something out of it. But here I want to be a bit like JF Kennedy. Don't ask what the energy company can do for you, ask what you can do for the energy company. We are trying to make the energy clients into good energy-system citizens.

This quote gives an idea of the logic employed by the utility when controlling the thermal environment in the home via the ADM system and the smart thermostats. This logic differed from the logic presented to participants in the experiment – of full control and, in a sense, liberation from the material constraints and 'normal' logic of the DH infrastructure. Households should, according to this logic, become optimal and efficient agents who serve the network. In other words, while homeowners are promised the use of an algorithmic 'servant' to care for their thermal comfort at home whenever and from wherever they wish, the algorithmic 'servant' is, in turn, expected to enable the users to work for the utility's interests.

The driving force here is the technical and economic optimisation of the energy system run by the utility, and this is achieved by knowing the user as well as possible (predicting demand), and overriding the temperature control in homes in a way that should be hardly perceptible to dwellers. Households give the utility access to the heating system at home and receive in return the promise of cleaner energy, free (or discounted) smart thermostats, and a guarantee that the temperature interventions by the utility will be imperceptible and will not compromise the sense of thermal home comfort. And if this is not enough, the user should at least try to be a 'good citizen', as the aforementioned quote requests, and yield to the demands of the energy company. The smart thermostat arrangement thus also incorporates a decision about the distribution of economic benefits: while shaving peaks saves costs for the utility and generates income for Ngenic, the user is asked to collaborate for the public good.

Trading data

The smart thermostats also connect the thermal comfort and economy of the home to a third realm, an economy of data. Data practices, automation and prediction have always played a vital role in the infrastructures of energy provision, as they have allowed utilities to craft a sense of seamlessness and uninterrupted provision of energy (Cohn, 2017). However, these data have until now been the domain solely of energy companies. ADM experiments such as Thermo-S allow companies from the data economy to insert themselves as an 'obligatory passage point' between energy companies and homeowners, gaining positions as data brokers. Such companies have only recently entered the 'world of energy', but they are changing the way energy utilities operate and how economic value is extracted from energy customers. As an Ngenic engineer argued,

> With our 25,000 clients today, we collect more data from them than all energy companies in Sweden do . . . and, new actors like us can look at the data with fresh eyes and build a completely new architecture.

Looking with 'fresh eyes' on data allows Ngenic to articulate it as a commodity and provide it as a service that is sold simultaneously to several markets, where homeowners, energy companies and data markets themselves are different 'clients': 'The source of data is the same. But we can use this data to provide a range of different services, both towards homeowners, and towards energy companies', Ngenic's engineer clarified. These services have been developed and tested as part of experiments such as Thermo-S that offer Ngenic a laboratory for trying different modes of using data and algorithmic steering in 'real-life' settings and fine tuning them towards different markets: 'When we don't guide the users towards the grid's purpose, then we guide them to optimize the building and the comfort in it. It is based on the same data. But it is just another service'.

In this new architecture, ADM itself becomes a commodity sold as a service, and utilities such as Jämtkraft become users to whom data from their own clients can be sold: 'They [the utility companies] have access to all this data. But we format them in a way that makes them useful for those running the engines, and for the distribution infrastructure', Ngenic's engineer explains.

Ngenic must ensure that users agree to share their data for multiple uses, to allow the company to carry out this formatting and create the new architecture. It achieves this by framing the sharing as an act of good citizenship. The company tells homeowners: 'Sharing your data makes you a good citizen. We will do the rest of the job. The only thing you need to do [as a user] is to say that this is OK'. The households are given the smart thermostat and in return they give away information about their energy use and indoor climate. This can then be processed by companies such as Ngenic to predict aggregate demand for the utility, or may be used by the ADM system to improve the quality of thermal services for homes.

Such data get also connected to the existing digital markets operated by data giants. For instance, the terms of use of the thermostats state that Ngenic's services share data with Google and Facebook for marketing and analysis and not just with energy companies (Ngenic – Data Protection, n.d.). By joining the Thermo-S experiment, users do not need to buy Google Nest: their data are seamlessly integrated as another stream into the digital infrastructures part of the platform economy (Gerlitz and Helmond, 2013).

Such an integration of users into global data markets is, however, never entirely smooth. Algorithms, and ADM systems, have obligations, and one of these is to produce satisfaction (Gillespie, 2014). Ngenic's algorithmic control is no exception – the premise of the smart thermostats and the data services they enable is to bring satisfaction to all involved parties in their distinct social worlds. However, these social worlds have contradicting understandings of the value of controlling temperatures – while users are promised an increase in their control of thermal comfort, the energy utility is promised the ability to control the users' flow of heat to a greater extent. This results in 'a conflict between the individual clients' interests and those of the system', as Ngenic puts it, and this is a conflict that the ADM system must accommodate. To resolve it, the algorithms in the ADM system are trained to weigh these different interests against each other, making sure that the satisfaction of the 'system' in terms of cost and energy efficiency is cared for, while 'No client should notice in terms of their thermal comfort when we exert control', Ngenic explains. Through this approach, using the results of the experiment to tune the ADM system determines power relations between users, the energy utility and Ngenic and inscribes particular hierarchies and priorities into the algorithms.

The value for Ngenic of experimenting with ADM in the Thermo-S project is then partly to see what can be done with the data in terms of packaging it as a service, and partly to test the socially and sensorially acceptable limits of controlling thermal comfort algorithmically, in order to make such multiple data service provision possible and seamless.

ADM and the algorithmic control of temperatures

The Thermo-S experiment connected and simultaneously managed three realms of actors and concerns. It created an environment for the thermal control of the home, where participants obtained enhanced possibilities to visualise and influence their in-house environment through a data platform on an app running on their phones. Such attempts to control thermal environments in buildings are not new. However, the tools through which they are implemented – thermostats, real-time data tracking and analysis, and ADM – convert mundane, everyday needs and desires about temperature regulation into an arena of algorithmic mediation where energy utilities, Big Tech and small tech actors try to capture new data markets and value.

The promise of better control and increased thermal comfort is a crucial argument for homeowners to participate in this experiment, although this promise is accompanied by an additional one: that these thermostats will help to run the whole DH system more efficiently and in an environmentally friendly manner. This promise links the comfort economy of the home to two additional realms of control that are driven by fundamentally different sets of values, interests and strategies: the social world of energy utilities with an interest in gaining control of the energy use in homes and managing this energy use in a way that helps to optimise the utility's energy system both economically and technically; the world of data and IT companies, who gain access to household energy use data that can, in turn, serve as the basis for new kinds of commercial service.

The achievement of the ADM system is to mediate between these three social realms and manage trade-offs in a way that sufficiently satisfies the different actors involved. On the one hand, the algorithm gives more user control, while, on the other hand, it reduces user control by allowing the energy company to override user temperature settings for the purpose of load management for as long as it remains hardly perceptible to the user. In reality, users never fully control the temperatures in their homes: the best they can hope for is that the mediated representations of temperature that they set in the app actually provoke some action, and that they sense a difference.

Moreover, the smart device enrols the household in a data economy interested in keeping users engaged on apps and generating data that are subsequently used in commercial services. The ADM system thus creates a new space for decision-making that did not exist before. While traditional thermal control systems are steered by a specific target temperature, decision-making in the Thermo-S system is much more complex and is driven also by load predictions (which in turn depend on data from other users), grid capacity and various economic considerations.

The decision-making algorithm of the ADM system is configured in a way that defines social relations between the actors involved (homes, utilities, data-processing companies) in terms of social power and economic gains. This experiment has tested the feasibility of this social and economic configuration and the compliance of users much more than it tested the technology: How much temperature variation do users accept, if they know it is of social and environmental benefit? How much data sharing do they tolerate? How large is the benefit of the data-based services provided to the utility?

Conclusions

The broader question that our analysis poses is, to what extent did the experiment make the 'reconfiguration of society' at stake here more transparent and politically negotiable. Sticking with Laurent's (2016) argument that an important function of an experiment is to create and convince publics, and that experiments are always at least partly uncertain and have an element of surprise, we can observe in our empirical case that these elements are unevenly distributed across the different actor worlds brought together by the smart thermostat. As Nadai and Labussiére (2018) point out, experiments often address different publics in different arenas simultaneously. The DH users as one of the publics addressed in this experiment are meant to be convinced that this experiment delivers better thermal comfort and control to the home, while there is at the same time no risk or uncertainty involved for the user. For the corporate publics of Jämtkraft and Ngenic, this experiment is rather a proof of concept which not only aims to demonstrate the technical feasibility of the system but also its ability to 'pacify' end-users, keep them satisfied and enrol them into this new configuration. At the same time, the experiment is aimed at a wider national public to which it demonstrates the innovativeness and boldness of Jämtkraft in taking first steps towards a new kind of energy system ('first digitalised district heating system in the world'). Despite these different publics and the narratives developed for them in the experiment, the new social and economic relations created through such experiments do not get much attention. The additional economic value created by the flexibility to manage heat loads and by the data syphoned off and used in commercial products appear to remain purely with Jämtkraft and Ngenic. Users give away both their 'flexibility capital' (Fjellså et al., 2021) and their data without getting much influence on their further use.

Questions can also be asked whether such systems reduce energy consumption and greenhouse gas emissions (as claimed by Jämtkraft) or whether the new control possibilities increase heat consumption by the need to rely, for instance, on computation taking place in data centres. Instead of laying bare the shifting economic and power relations between the actors of the new smart energy system tested in this experiment, the ADM algorithm renders all these relations invisible, by creating an illusion of autonomous control in homes while at the same time configuring decision processes in ways that benefit the commercial interests of the energy suppliers and data companies that provide new data services. Customers who are not connected to a smart device can always change the setting of the radiators manually to save energy, but passing control to the energy company may, paradoxically, lead to the consumer taking fewer active choices and forgetting that active choices are possible.

The broader relevance of such experiments that examine the algorithmic mediation between infrastructure providers, data-driven tech companies and infrastructure users should not be underestimated. While energy infrastructures in the past were driven by an ideal of providing universal service, capacity reserves and a network expansion that followed growing demand, this has been increasingly replaced by attempts to handle capacity problems by 'managing users', by collecting and processing data about their behaviour, and the subsequent control of their infrastructure use by economic incentives, appeals to citizenship, and incremental manipulation of the infrastructure-based services they receive. These new arrangements apply to all kinds of infrastructure, such as heat and electricity networks, electric vehicle charging facilities and traffic management. What is tested here is indeed a 'reconfiguration of society', and the analysis of such experiments as we have attempted in this chapter may help to re-politicise questions of the distribution of social power and economic benefits, privacy and data sovereignty, transparency of control, the socially uneven access to new infrastructure services, and the individualisation of environmental responsibility.

References

Ackermann ME (2010) *Cool Comfort: America's Romance With Air-Conditioning*. Washington, DC: Smithsonian Books.

Ansell CK and Bartenberger M (2016) Varieties of Experimentalism. *Ecological Economics* 130: 64–73. https://doi.org/10.1016/j.ecolecon.2016.05.016

Benson-Allott CA (2015) *Remote Control: Object Lessons*. New York: Bloomsbury.

Beregow E (2019) Editorial: Thermal Objects: Theorizing Temperatures and the Social. *Culture Machine* 17.

Cohn J (2017) Data, Power, and Conservation: The Early Turn to Information Technologies to Manage Energy Resources. *Information & Culture: A Journal of History* 52(3): 334–61. https://doi.org/10.1353/lac.2017.0013

Cooper G (2002) *Air-Conditioning America: Engineers and the Controlled Environment, 1900–1960*. Johns Hopkins Studies in the History of Technology N.S., 23. Baltimore, MD: Johns Hopkins University Press.

Engels F, Wentland A and Pfotenhauer SM (2019) Testing Future Societies? Developing a Framework for Test Beds and Living Labs as Instruments of Innovation Governance. *Research Policy* 48(9): 103826. https://doi.org/10.1016/j.respol.2019.103826

Fjellså IF, Silvast A and Skjølsvold TM (2021) Justice Aspects of Flexible Household Electricity Consumption in Future Smart Energy Systems. *Environmental Innovation and Societal Transitions* 38: 98–109. https://doi.org/10.1016/j.eist.2020.11.002.

Gabrys J (2014) Programming Environments: Environmentality and Citizen Sensing in the Smart City. *Environment and Planning D: Society and Space* 32(1): 30–48. https://doi.org/10.1068/d16812

Gerlitz C and Helmond A (2013) The Like Economy: Social Buttons and the Data-Intensive Web. *New Media & Society* 15(8): 1348–65. https://doi.org/10.1177/1461444812472322

Gillespie T (2014) The Relevance of Algorithms. In: Gillespie T, Boczkowski P and Foot K (eds) *Media Technologies: Essays on Communication, Materiality, and Society*. Cambridge: MIT Press, 167–94.

Google Nest Help (n.d.) How a Nest Thermostat Helps Save Energy. Available at: https://support.google.com/googlenest/answer/9254386?hl=en-GB (accessed 30 April 2021).

Gram-Hanssen K (2010) Residential Heat Comfort Practices: Understanding Users. *Building Research & Information* 38(2): 175–86. https://doi.org/10.1080/09613210903541527

Jämtkraft (2018) Åre först i världen med digitaliserat fjärrvärmenät. Available at: www.jamtkraft.se/om-jamtkraft/nyhetsrum/are-forst-i-varlden-med-digitaliserat-fjarrvarmenat/ (accessed 30 April 2021).

Laurent B (2016) Political Experiments That Matter: Ordering Democracy from Experimental Sites. *Social Studies of Science* 46(5): 773–94. https://doi.org/10.1177/0306312716668587

Light B, Burgess J and Duguay S (2018) The Walkthrough Method: An Approach to the Study of Apps. *New Media & Society* 20(3): 881–900. https://doi.org/10.1177/1461444816675438

Marres N (2020) Co-existence or Displacement: Do Street Trials of Intelligent Vehicles Test Society? *The British Journal of Sociology* 71(3): 537–55. https://doi.org/10.1111/1468-4446.12730

Marres N and Stark D (2020) Put to the Test: For a New Sociology of Testing. *The British Journal of Sociology* 71(3): 423–43. https://doi.org/10.1111/1468-4446.12746

Nadai A and Labussiére O (2018) Technological Demonstration at the Core of the Energy Transition. In: Nadai A and Labussiére O (eds) *Energy Transitions. A Socio-technical Inquiry.* Cham: Palgrave Macmillan, 191–237.

Ngenic (n.d.) Ngenic Tune. Available at: https://ngenic.se/en/tune/ (accessed 10 May 2021).

Ngenic – Data Protection (n.d.) Dina personuppgifter och hur vi hanterar dem. Available at: https://ngenic.se/dataskydd/personuppgifter/ (accessed 12 May 2021).

Ngenic – Terms of Service (n.d.) Användarvillkor. Available at: https://ngenic.se/dataskydd/anvandarvillkor/ (accessed 26 August 2021).

Shove E (2003) *Comfort, Cleanliness and Convenience: The Social Organization of Normality.* London: Bloomsbury Academic.

Strengers Y (2013) *Smart Energy Technologies in Everyday Life. Smart Utopia?* Basingstoke: Palgrave Macmillan.

Suchman LA (2007) *Human-Machine Reconfigurations: Plans and Situated Actions.* Cambridge: Cambridge University Press.

Thrift N and French S (2002) The Automatic Production of Space. *Transactions of the Institute of British Geographers* 27(3): 309–35. https://doi.org/10.1111/1475-5661.00057

12

PRISONERS TRAINING AI

Ghosts, humans and values in data labour

Tuukka Lehtiniemi and Minna Ruckenstein

Introduction

As we enter the open-plan office of Vainu, a Finnish start-up based in a converted industrial building in Helsinki's gentrified Kallio neighbourhood, a large screen displaying charts greets us from the wall. The company representative tells us that we are seeing statistics on their use of on-demand data labourers, who produce training data for Vainu by skimming textual data and labelling words that refer to named entities such as organisations. In doing so, they are training Vainu's artificial intelligence (AI). Vainu, which takes its name from the Finnish word for 'hunch', sells data and analytics to other companies for the purpose of improving business-to-business sales. The data originate from public sources: financial statements, articles in the business press, social media and website content. Vainu's machine learning models automatically generate structured data from these unstructured data sources, and the on-demand data labour displayed on the screen in the office comes into play when the models encounter something the machine does not recognise. At that point, more training data are automatically queried through application programming interfaces (APIs) from on-demand platforms like Amazon Mechanical Turk (MTurk).

The screen visualising Vainu's use of data labour speaks to concerns of a rapidly expanding realm of research, covering issues that have to do with automation and digital platforms (Crawford, 2021; Gray and Suri, 2019; Mateescu and Elish, 2019; Newlands, 2021; Vallas and Schor, 2020). In this chapter, we are particularly interested in the organising of human work that supports, and becomes an integral part of, automated processes. We build on interdisciplinary research, ranging from sociology and critical data studies to anthropology and media research, and emphasise human involvement in data production and automation more generally. In studies that examine datafication as a societal process underlying new forms of automated decision-making (ADM), human participation, including how humans

DOI: 10.4324/9781003170884-16

enable automation by doing machine-like, standardised work that machines cannot perform, tends to disappear from sight (Crawford, 2021; Mateescu and Elish, 2019). This disappearance is at least partially a result of purposefully managed invisibility: AI providers strategically obscure and occlude human efforts that go into the creation and training of their services (Newlands, 2021). If human work in data is accounted for in research at all, the focus is typically on processes of organising, analysing and judging that add value (e.g. Foster et al., 2018). Human involvement, then, refers to the expert skills of converting data into knowledge: the work of data scientists and data visualisers. Nevertheless, data rarely exist in a usable format even for experts or the forms of automation they promote without human labour being first involved in turning data into a usable resource. Here, the data work that precedes data analysis is crucial: efforts that go into cleaning and editing data, generating metadata such as labels and annotating data with contextual information (Møller et al., 2020).

In this chapter, we discuss ADM by way of 'the hidden faces of automation' (Irani, 2016): data labour that keeps the infrastructures of the data economy and its automated functions running. Mary Gray and Siddharth Suri (2019) call such data labour 'ghost work' due to its unheralded and largely unnoticed nature. We use as our example of ghost work a data labour arrangement in which prisoners label Finnish language data for a local AI firm. We base our observations on interviews with representatives of the AI firm Vainu and prison officials and on documentary material on the prison data labour project. Labelling text to produce metadata for training an AI system is a typical example of ghost work: machine-like on-demand work that cannot be performed by machines, at least without initial human input. Prisoners in closed prisons are a marginal group that by definition tends to remain hidden – when they perform on-demand data labour, its ghost-like qualities become accentuated. As well as noting similarities with other types of ghost work, we show that the Finnish prison is an unconventional and ambiguous site for data labour, raising questions about low-tech workers performing high-tech work in a digitally deprived environment.

In existing research, data labour is often viewed in dystopian terms. Platforms are seen to accelerate precarity and inequality, and they are described as digital cages that disenfranchise workers (Vallas and Schor, 2020). As we demonstrate, however, Vainu's use of prison data labour opens other possible ways to think about ghost work and questions of value extraction. We treat prison data labour as a fringe case that highlights ghost work in ways that might not become visible otherwise and thereby allows us to engage with alternative visions of ADM. The Finnish prison data labour arrangement draws attention to collaborations, tensions and ambiguities around data-based automation, calling for critical inquiry that is able to hold seemingly contradictory aspects together without resolving them into a totalising perspective or a straightforward story of exploitation. In demonstrating different aspects of the prison data labour case, we build on a processual understanding where different values might be at stake for the collaborating parties. In this view, value is located not in aesthetically desirable objects, monetary transactions or people but in how relations to technological processes, including various kinds of agencies, are mediated (Graeber, 2001). According to David Graeber (2001), value is related to actions, emergent in social aims, and within particular sets of circumstances. Our prison data labour exploration pays attention to the different alignments and agencies at play. It demonstrates what is of value to the different parties involved in the organisational arrangements of AI training, as they both work with and intervene in political–economic incentives and pressures, such as platformisation and automation. The fact that prison data labour appears as a win–win project for all parties involved suggests that it successfully combines AI-related aspirations, offering a multi-stakeholder view of the current ghost work landscape.

We will demonstrate how the case of prisoners training AI highlights the local and situational variations of human labour underlying ADM. Here we emphasise the many moving parts involved in arrangements that allow data labour and ADM to come into being. What emerges as significant is the anticipatory nature of data labour, which is hardly coincidental given how digital technologies intertwine with prospective futures. More surprising, perhaps, is that anticipation is geared towards improving the lives of the most marginal inmates in Finnish closed prisons. Caring for what technologies do for the most vulnerable, whether in prison or elsewhere, accentuates the need to explore what technologies promote in the everyday, by whom and for whom, and based on what assumptions and value orientations. The case demonstrates that novel ways of thinking about processes of ADM can be opened up by tracing what is of value to the different parties involved and how their value aims are – or could be – combined in practice.

Bringing data labour to prison

In March 2019, Vainu (2019) announced that AI had entered Finnish prisons and that inmates were now participating in the development of Finnish language AI. The unusual collaboration with the Finnish Criminal Sanctions Agency (known as RISE, its Finnish acronym) was set in motion by Tuomas Rasila, a serial entrepreneur and one of Vainu's founders, and covered in the international media (Chen, 2019; Peteranderl, 2019). An article on *The Verge* (Chen, 2019) raised the question of whether prison data labour is empowering or exploitative but did not take an explicit stand on that question, noting that the local contexts of prison labour vary. In the article, Lilly Irani, an associate professor specialising in cultural politics of high-tech work, points out the marketing angle of the project: 'this kind of hype circulating around AI . . . can masquerade really old forms of labour exploitation as "reforming prisons"'. Indeed, prisons are characterised by the vulnerable position of prisoners, highly unequal power relations, extreme control and surveillance and forms of coercion that are otherwise practically non-existent and largely unimaginable in democratic societies.

Rasila, the entrepreneur behind the prison data labour project, however, thinks positively about the problem that he aims to solve rather than its societal implications. With previous experience of technology projects, he started thinking that with the proper arrangements in place, prisoners might help train AI to process company-related materials. A prison is, at least when viewed from the outside, a place where people have a lot of time in their hands. Moreover, the digital backwardness of prisons calls for technical innovation (Kaun and Stiernstedt, 2021). Rasila tells how he develops ideas by making extremes meet; this is, for him, where new and interesting things tend to happen. In an AI training project situated in a prison, progress could be married with the marginalised. As Rasila puts it: 'The world's most advanced AI meets an inmate who is in some sense outside of society. And as a result, we get something extremely sophisticated . . . an unwanted, low-end workforce can participate in high-end development.'

Rasila's thinking accentuates the local and situated nature of automation projects. A company like Vainu needs to deal with data that require location- or culture-specific human interpretation, such as signalling to the machine whether a named entity in a Finnish language text is a municipality, a company or a mobile phone, as in the case of 'Nokia'. A workforce with the necessary local skills is simply not available on regular platforms like MTurk, at least on the scale that Vainu needs, because a Nordic welfare state provides social benefits that make low-paying data labour unattractive. Automatic translation of Finnish texts to English does not help either, because it tends to produce too many errors and misinterpretations.

Previously, Vainu had tasked interns to generate the necessary Finnish language data to train AI. The company had also considered hiring data labour via staffing agencies and had made use of platforms such as UpWork in other small-language markets. These approaches, we were told, were incompatible with the data needs of machine learning systems. Even if hiring contingent staff or freelancers for data labour would be more costly than using MTurk, the issue is not so much cost as the nature of the demand. What Vainu needs is a constant availability of many workers to produce the appropriate amount of training data in a timely manner, not a sporadic, temporary workforce doing the work in batches. It is the flexibility and on-demand nature of data labour that most interested the company. From Vainu's standpoint, the need for data labour is not diminishing; on the contrary, even if machine learning models learn to solve specific problems, other problems will emerge that necessitate new training data.

RISE officials thought favourably about the prison data labour project. Collaborations between prisons and companies have gradually decreased due to prisons losing their competitive edge to cheap labour outsourced from the Global South, which makes new openings for prison work attractive in general (Kaun and Stiernstedt, 2020). Moreover, for reasons that are clarified later, a pilot was quickly set up. AI training took place over a 15-month period in two closed prisons in Finland, one in Helsinki and the other in Turku. The pilot was, at the time, viewed by both Vainu and RISE as the initial stage of a longer-term collaboration. In practice, data labour was introduced in three closed prison wards. A few dozen prisoners trained the AI by using customised laptops running purpose-built software to highlight words referring to different types of named entities with different colours. The computers and the software provided by Vainu were audited for security purposes. Prisoners worked under the watchful eyes of prison officers in spaces allocated for the purpose, even though it was stressed that the work was flexible in terms of space, and in later stages of the project, the work could even be moved to prison cells.

Vainu's representatives were pleasantly surprised by the quality of the work done by inmates: it was higher than they had learned to expect from the work on MTurk. The company carefully stressed in its communications that they were only interested in the final outcome – annotated data for AI training – and were not aware of or even interested in arrangements inside the prison. The organisation of data labour typically hides the individual worker: data labourers on ghost work platforms are ID numbers and codes, not identifiable by name or location. A requester of work specifies eligibility criteria for tasks rather than specifying who does the work (Gray and Suri, 2019). While Vainu's system could connect tasks to pseudonymous IDs, the company had no control over or even awareness of who logged in with a particular ID.

Overall, Rasila considered the prison data labour pilot to be a test or proof of concept. It showed that it was possible for Vainu to have practically anyone willing to do data labour work for them and to automatically audit the quality of the result. Rasila and his colleagues had even semi-seriously entertained the idea of a more general platform that would intermediate data labour tasks to different locations where people have extra time. Alas, this was not in Vainu's business interests, but it proved the point: if data labour could be done in prisons, it could be done anywhere. In this respect, prison operates as an accelerator of automation processes, a test bed for technological development, exploring concepts and prototypes that can later be transitioned from the prison into other use areas (Kaun and Stiernstedt, 2021).

The value of prison data labour

Globally and historically speaking, prison labour is typically punitive by nature, whether it is productive, pointless hard work or serves more indirect institutional needs (Kaun and Stiernstedt, 2020). The abusive potentials invite a straightforward conclusion that prison data labour is yet another form of punishment. When productive, prison labour can appear to be a straightforward exploitation of the prisoners' bodily or mental capabilities for the benefit of others. Moreover, prison data labour is an example of digital production, which is tied to other kinds of exploitative potentials (Andrejevic, 2008; Bruns, 2008; Scholz, 2013; Terranova, 2000). Yet, within the Finnish penal system, prison data labour tries to boost the agency of prisoners in ways that aim to alleviate inequalities, making it a value project for the prison system.

Finland, like the Nordic countries in general, maintains a 'rational and humane criminal policy' which has its roots in the welfare state with its consensual political culture and high levels of public trust (Lappi-Seppälä, 2007: 219). The Finnish penal policy has a human rights focus, in that prisoners are not 'slaves of the state'; their rights are protected similarly to other citizens (Lappi-Seppälä and Nuotio, 2019). Social and criminal policies are also intimately intertwined; welfare provisions as measures against social marginalisation and inequality are also measures against crime (Lappi-Seppälä and Nuotio, 2019). When punishment is issued, alternatives to imprisonment such as suspended sentences and community sanctions are predominantly used. Due to these policies, incarceration rates in Finland are low by international standards. In 2019, Finnish prisons held a total of 2,748 inmates, which translates to an incarceration rate of 50 inmates per 100,000 inhabitants – around half the European median (Aebi and Tiago, 2020), about 25% of the Australian rate (Australian Bureau of Statistics, 2021) and about 7% of the US rate (Prison Policy Initiative, 2020).

The data labour pilot took place in closed prison wards. To be incarcerated in a closed prison, inmates must either be repeat offenders or have committed an exceptionally grievous offence. The Finnish penal code (Ministry of Justice, 2005: Section 3) states an intent to maintain the health and functional abilities of those in closed institutions so that detrimental outcomes are minimised. The loss of liberty is considered a punishment in itself, and the imposition of additional hardship is neither required nor permitted. In fact, the aim of imprisonment is to increase the ability of convicted criminals to readjust to society and lead a crime-free life. The 'normality principle' expressed in the Finnish penal code asserts that conditions in prison should correspond, as closely as possible, to living conditions outside the walls (Lappi-Seppälä and Nuotio, 2019).

Our aim is not to paint too rosy or idealistic a picture of Finnish prisons. Based on earlier ethnographic experience in a closed prison, prisons in Finland are also harsh and hopeless places (Ruckenstein and Teppo, 2005). Despite the human rights focus in Finland's penal policy, the anti-torture committee of the Council of Europe has long expressed concern about primitive cells in the oldest Finnish prisons and has more recently noted the inadequate facilities and conditions to which segregated inmates are subjected (CPT, 2021). Nevertheless, in light of the Finnish penal ideology – minimisation of harm, normality and a focus on adjustment to society – careful consideration is given to aspects of prison labour that are considered helpful in terms of eventual resettlement. In this context, prison data labour tries to facilitate inmates' return to civilian life.

Before turning to the question of how prison data labour might align different value aims, and be beneficial to the prisoner, however, the price of data labour merits consideration. As part of the pilot project, Vainu paid RISE for data labour. The pricing arrangement is similar to those typically employed on MTurk and other ghost work platforms: Vainu pays by the task within the range of the market value of click work tasks. The median overall pay on MTurk is two US dollars per hour (Hara et al., 2018). The price that Vainu pays to workers on MTurk is not among the lowest: according to a task listed in late 2020, for example, Vainu offered a reward of US$0.36 for labelling references to organisations, persons, locations and job titles in a business news article. In Finnish terms, however, this remains extraordinarily low compensation.

The price per task RISE defined for Vainu was calculated with the same system that evaluates the price of any prison labour: an estimation of the number of tasks prisoners are expected to perform in a specified timeframe. Prisoners, however, are not compensated based on tasks but receive a modest allowance that is not exactly a salary. Prisoners in closed wards are expected to participate in freely chosen 'prison activities': prison labour, courses, training and rehabilitation programmes. The daily allowance for participating in activities is defined by law. Of the two possible allowance levels, data labour provides the slightly higher one, amounting to 4.62 euros per day, which means that inmates benefit from choosing it. Prison activity does not mean a full working day or meeting a specified quota. Typically, prisoners do not work more than a few hours, and this was also the case with the data labour for Vainu.

Setting the economic value for prison labour underlines the local rootedness of the case: however the work is organised – and whatever its relation to punishment or rehabilitation – Finnish prison data labour remains exceptional internationally. It builds on the local history of prison labour but is also in dialogue with the larger tendencies of automation, particularly the need to grow masses of cheap data labourers. The prison appears to be a promising frontier for expansion: cheap in comparison to the alternatives available for culture- or language-specific data labour in the Nordics and thus far untouched by competitors. While the data labour arrangements are not as nefarious as some other technologies used for organising prisons, such as those for surveillance or individualised control (Foucault, 2012; Rhodes, 2004), forms of data labour can nevertheless be harmful. Content moderation involving the assessment of violent and hateful content is one example of psychologically damaging data labour (Roberts, 2019). The Finnish prison authorities consider the suitability of prison data labour in terms of safety for inmates, societal consequences and the moral values that the work embodies. Having prisoners shift through violent and pornographic content would, of course, be unthinkable. In the case of Vainu's data labour, RISE saw no potential issues. The material that the inmates scan – news articles and other texts – were considered harmless to work with, as were the tools for the work: computers, unlike hammers or screwdrivers, cannot be used to do physical harm, at least not in such obvious ways. The purpose of the AI being trained – ultimately, more efficient sales operations – was also considered acceptable.

Aspirational data labour

While those on the outside are likely to consider data labelling as simple and menial, the prison officials considered reading and annotating to be cognitively challenging and more rewarding than the alternatives available to prisoners. Our informants at RISE emphasised that many inmates in closed prison wards struggle with learning; as a result, their self-esteem as learners can be low. Since data labour requires skills like reading and interpreting text, it introduces knowledge-oriented work into a closed prison environment. Typical forms of prison labour, such as putting screws in boxes, allow you to 'check your brain at the door', as a prison counsellor puts it. By contrast, data labour obliges prisoners to be more mentally active, as the work involves problem-solving and requires close attention.

In the context of the prison, data labour becomes constructed as a site of anticipation 'that sets the conditions of possibility for action in the present' (Adams et al., 2009: 249). It is seen as beneficial in terms of aligning prisoners with the knowledge society and developing and maintaining their capabilities. Rasila connects data labour to an intrinsic motivation that he expects inmates to experience:

If I consider what inspires me personally, it's when I know I'm doing something useful for others. In work like [the annotation work that prisoners do] this happens right away, because the tasks can be done so quickly. When a task is done, it has resulted in a better version of the AI that saves human effort in a permanent way. Things have clearly moved forward. I believe it can be an empowering feeling for a person who has in a sense fallen outside of society, when they can help the society by doing work like this.

The assumption here is that inmates aspire to make a meaningful contribution, and high-end technology appears as a means to make that contribution possible, even from the very margins of the society. The inmate is expected to be motivated to move towards society, even though in reality prisoners in closed prisons might have long ago given up on such desires. Overall, data labour becomes associated with a curious form of techno-optimism, rarely found in the prison, where technology is typically an apparatus of control and surveillance rather than individual opportunity.

The hopes and expectations that AI training upholds reveal a strongly aspirational element in prison data labour. Inside the prison, the anticipation that data labour is valuable for prisoners materialises in the choice of atypical prison wards. RISE appeared to treat data labour as an opportunity to offer a new type of work to prisoners who are in a particularly vulnerable position in the prison, and data labour was first introduced in a female ward in an otherwise male prison. If prisoners in general are marginal in Finnish society, female prisoners are on the margins of the already marginal. The project manager at RISE commented that female prisoners could for once be part of something positive: 'It is the intelligence of Finnish female prisoners that guides Vainu's system to deliver information to companies and customers', she said, underlining the valuable contribution that women's cognitive efforts could make for AI. In addition to female prisoners, youngsters in mostly adult prisons and prisoners serving time in solitary confinement (often at their own request, as in the case of sex offenders) were deemed by RISE to be suitable candidates. As some traditional forms of prison labour, such as woodwork and metalwork, require dedicated workshop spaces and are done in groups, certain inmate groups cannot participate in these activities. The more difficulties inmates have fitting in with the prison population, the more they benefit from the flexibility and independent nature of data labour. Thus, data labour is seen as a particularly promising area of rehabilitative work: at least in theory, it is safe and doable in closed wards and even solitary cells.

The promoters of the pilot project, then, hoped that training AI would provide inmates with self-confidence in an increasingly digitalised job market – and a digitalised society more generally. Despite attempts to build smarter facilities, prisons have largely remained a digitally deprived environment, making prisoners an impoverished group in the digital age (Reisdorf and Jewkes, 2016). Prisoners in Finland's closed prisons have extremely restricted access to computers and information networks. In the prison context, computers are so out of place that they can be viewed as highly charged and suspicious objects. This kind of suspicion extended to the computers used in the data labour pilot. As one of the prison counsellors told us, the assumption might be that once a prisoner gets access to a computer, they 'go to the dark web to buy drugs'. Despite reservations about digital tools, however, the normality principle outlined in the Finnish penal code – that conditions in prison should be as close to outside society as possible – means that prisons cannot remain a digitally deprived island. As one prison official stressed, 'It is not possible to participate in society without access to digital services and the capabilities to use them'. Accordingly, the Finnish prison authorities are planning to equip prison cells with computer terminals to be used for e-services and communication (Järveläinen and Rantanen, 2020). Computerised prison labour that could be performed on these terminals fits in with the ongoing promotion of prisoners' digital inclusion.

It was striking how the possibility that prisoners could benefit from data labour energised our informants. Their future speculations were not only geared towards technical advances, as is typical for enthusiastic stories of ADM, but they also stayed close to how they might help the prisoner. In the prison world, our informants at RISE stressed, success comes in very small steps. Despite the aims of a humane penal policy, a closed prison is still a black hole that can suck the air out of all positive future scenarios. Any kind of anticipation that points towards progress or improvement in the lives of prisoners is welcome. A prison counsellor who oversaw female prisoners' data labour told us about their concentration, sitting silently in a shared space with their computers, completing their tasks. At times, the women would ask each other how to solve a problem. With these observations, he emphasises that prisoners are fully engaged in what they are doing – something that cannot be taken for granted. We heard more than once an anecdote about a female prisoner belonging to a marginal ethnic minority in Finland who had no experience with computers. The goal of this story was to concretise the many benefits of the prison project. After being encouraged to try data labour, the woman said that she enjoyed reading the news items that the system presented to her. Reading texts that she would never have otherwise encountered moved her in a new direction, and she became more comfortable with computers. She also liked being able to say that she had participated in creating AI while serving her time; it sounded much better than folding socks or packing screws.

For Brooke Erin Duffy (2017), aspirational labour covers productive activities in the digital environment that participants believe will bring a future payoff in the form of material rewards, social and economic capital or professional opportunities. The temporal element is central to aspirations, in that the value and the rewards from the work done are expected to materialise in the future. Working now is treated as an investment with future returns. Duffy's aspirational labourers are female content creators in the digital culture industry, with their mostly unpaid activities propelled by the ideal of getting paid to do what you love at some point. In the case of prison data labour, we see a similar shift towards expected future value: work is imagined to be beneficial in the long run, as it aids in the eventual resettlement.

Importantly, however, the future returns of prison labour are not necessarily or mainly the inmates' own aspirations. They might well be, but what we observed – and what formed the entire basis of the prison pilot – were others' rehabilitative aspirations on the prisoners' behalf, based on a professional evaluation of what would be needed after resettlement. In that sense, the aspirations are more mundane and down to earth than Duffy's digital content creators' hopes of getting economic and social rewards for what pleases them. The aspiration is that the inmates, with very little to expect from life in terms of work, would have slightly better opportunities to become functioning participants in the increasingly digitalising society. The fact that AI training is cheap labour that is fragmented, unbundled from context and bereft of prospects for professional development (Crawford, 2021) is rendered insignificant when aspirations do not concern the future returns from the work itself but rather the opportunities it offers for healing and readjusting to society.

Humans, not ghosts

As we followed the data labour project, both Vainu and the prison administration expressed satisfaction with its results. The project was set to be continued: our informants cited plans to expand beyond the prison wards selected for the pilots, and workers in new prisons were trained as supervisors. Just as we had secured research permits to carry out observations and interviews inside prisons, COVID-19 closed the prison gates to all visitors. When restrictions were about to be lifted in the summer of 2020, in a move that took both of us and the prison administration by surprise, Vainu decided to end the project. That decision was made by a new management team that had no first-hand experience with or appreciation of the project. The civil servants at RISE tried to convince Vainu to continue the project, even referring to our research proposal to underline the value of the collaboration. Vainu had, however, already decided that the prison project was not part of their core business and wanted to move on without it.

Our informants had a hard time accepting that a project in which they had invested so much time and effort and that appeared to be a win–win proposition in many ways, supporting both the prison system's aims and the company's need for cheap labour, could simply be terminated. The desire to hold on to the project accentuated that the partners involved in the project felt that they had successfully aligned societal and economic value aims. The failure of the new management to appreciate the project, in fact, meant that they had failed to understand and mobilise the value which emerged from this particular set of circumstances. For us, the abrupt ending of the project underlined the short-sighted interests of private entities in crafting the futures of ADM. While companies eagerly start new initiatives with the support of public sector resources, their commitments can remain uncertain and their actions unaccountable. Most companies appear to have little imagination when it comes to the potential futures of automation; they follow the crowd and do what others have done before them. With this attitude, it is not possible to commit to initiatives that seek to generate new kinds of value with unlikely collaborations and do something unprecedented. However, it might require precisely the unprecedented in order to successfully automate what appear as rather ordinary-seeming tasks. Finnish language training data are not readily available on regular ghost work platforms that provide on-demand data labour with only a specific skill profile.

Paradoxically, on these platforms, global income inequality dictates the availability of training data and the tasks that can and cannot be automated.

Given its abrupt ending, the prison data labour pilot ended up being a short-lived experiment, which did not provide a clear resolution on the merits of prison data labour. It still generated something productive (Marres and Stark, 2020): for a brief moment, it demonstrated an adventurous combination of different aspirations and value aims, while a public–private partnership successfully mediated between the potentially incompatible aims of the collaborating parties. This is precisely why it was described as a win–win project. With fitting collaborations, the pilot could have been expanded into a more general-purpose data labour platform that could cater to additional data needs. When considering such a platform, hypothetical as it may be, our case draws attention to how mistaken it would be to make straightforward analogies between Vainu's prison data labour arrangement and a regular ghost work platform. MTurk, for instance, acts as an intermediary in two senses: it offers a pathway for machines to access humans that perform data labour and it matches data labourers with tasks on offer. When Vainu built a data labour pipeline that allowed the prisoners to perform data annotation tasks, it also performed the role of an intermediary. Yet, this intermediation was strikingly different from MTurk. Vainu built the technology that ensured the transfer of annotation tasks between the machine and human ends of the pipeline, but it was the prison and its workers that ensured that inmates were available to carry out the tasks. Whereas ghost work platforms hide data labourers, inmates are visible to the prison officials and counsellors considering the suitability and rehabilitative potential of data labour. In the prison, the matching of humans and machines takes into consideration the prison system's value-related aims in relation to prisoners, creating new kinds of caring relations in the process. This means that inmates are not ghosts but real people in everyday life conditions that leave much to be desired.

The prison data labour case underlines that any arrangements around ADM depend on local and contextual relationalities and variations. These aspects need to be carefully considered so that we do not lose important differences and end up seeing only techno-deterministic futures. Even if data labour might be considered unappealing, exploitative or even dehumanising, these qualities become sidelined when the goal is to improve the lives of those in the margins of the digitalising society. Not only is AI training by Finnish prisoners firmly embedded in existing inequalities, but it also tries to work with them, reminding us of how the past, present and future simultaneously influence the field of ADM. On a closer look, the aims and values involved in prison data labour depend on the prevailing penal philosophies and practices, including the penological function of prison labour in preventing recidivism and repressing criminal behaviour. Yet, perhaps even more importantly, the case highlights the role of humans in processes that turn data into a resource that can be built on, drawing attention to human involvements and imaginaries that are crucial in promoting automated futures. Such involvements demonstrate anticipations, collaborations and eventual disconnects around data-based automation, making it plain that the humans, with their guiding values, are the most critical component in human–machine arrangements.

References

Adams V, Murphy M and Clarke AE (2009) Anticipation: Technoscience, Life, Affect, Temporality. *Subjectivity* 28(1): 246–65.

Aebi MF and Tiago MM (2020) *SPACE I – 2019 – Council of Europe Annual Penal Statistics: Prison Populations*. Strasbourg: Council of Europe.

Andrejevic M (2008) Watching Television without Pity: The Productivity of Online Fans. *Television & New Media* 9(1): 24–46.

Australian Bureau of Statistics (2021) Corrective Services. Available at: www.abs.gov.au/statistics/people/crime-and-justice/corrective-services-australia/dec-quarter-2020 (accessed 7 September 2021).

Bruns A (2008) *Blogs, Wikipedia, Second Life, and Beyond: From Production to Produsage*. New York: Peter Lang.

Chen A (2019) Inmates in Finland Are Training AI as Part of Prison Labor. *The Verge*, 28 March. Available at: www.theverge.com/2019/3/28/18285572/prison-labor-finland-artificial-intelligence-data-tagging-vainu (accessed 7 September 2021).

CPT (2020) *The CPT and Finland*. Available at: www.coe.int/en/web/cpt/finland (accessed 7 September 2021).

Crawford K (2021) *Atlas of AI. Power, Politics and the Planetary Costs of Artificial Intelligence*. New Haven, CT: Yale University Press.

Duffy BE (2017) *(Not) Getting Paid to Do What You Love. Gender, Social Media, and Aspirational Work*. New Haven, CT: Yale University Press.

Foster J, McLeod J, Nolin J and Greifeneder E (2018) Data Work in Context: Value, Risks, and Governance. *Journal of the Association for Information Science and Technology* 69(12): 1414–27.

Foucault M (2012) *Discipline and Punish: The Birth of the Prison*. New York: Vintage.

Graeber D (2001) *Toward an Anthropological Theory of Value. The False Coin of Our Own Dreams*. New York: Palgrave Macmillan.

Gray ML and Suri S (2019) *Ghost Work. How to Stop Silicon Valley From Building a New Global Underclass*. Boston, MA: Houghton Mifflin Harcourt.

Hara K, Adams A, Milland K, Savage S, Callison-Burch C and Bigham JP (2018) A Data-Driven Analysis of Workers' Earnings on Amazon Mechanical Turk. *Proceedings of the 2018 CHI Conference on Human Factors in Computing Systems*, Montréal, Canada, 21–26 April. New York: ACM.

Irani L (2016) The Hidden Faces of Automation. *XRDS* 23(2): 34–7.

Järveläinen E and Rantanen T (2020) Incarcerated People's Challenges for Digital Inclusion in Finnish Prisons. *Nordic Journal of Criminology*. Epub ahead of print 14 September. https://doi.org/10.1080/2578983X.2020.1819092

Kaun A and Stiernstedt F (2020) Prison Media Work: From Manual Labor to the Work of Being Tracked. *Media, Culture & Society* 42(7–8): 1277–92.

Kaun A and Stiernstedt F (2021) Prison Tech: Imagining the Prison as Lagging Behind and as a Test Bed for Technology Advancement. *Communication, Culture and Critique*. Epub ahead of print 25 June. https://doi.org/10.1093/ccc/tcab032

Lappi-Seppälä T (2007) Penal Policy in Scandinavia. *Crime and Justice* 36(1): 217–95.

Lappi-Seppälä T and Nuotio K (2019) Crime and Punishment. In: Letto-Vanamo P, Tamm D and Gram Mortensen B (eds) *Nordic Law in European Context*. New York: Springer, 179–99.

Marres N and Stark D (2020) Put to the Test: For a New Sociology of Testing. *British Journal of Sociology* 71(3): 423–43.

Mateescu A and Elish MC (2019) *AI in Context: The Labor of Integrating New Technologies*. New York: Data & Society.

Ministry of Justice (2005) *Imprisonment Act* 23.9.2005/767. Available at: https://finlex.fi/en/laki/kaannokset/2005/en20050767 (accessed 7 September 2021).

Møller NH, Bossen C, Pine KH, Nielsen TR and Neff G (2020) Who Does the Work of Data? *Interactions* 27(3): 52–5.

Newlands G (2021) Lifting the Curtain: Strategic Visibility of Human Labour in AI-as-a-Service. *Big Data & Society* 8(1): 1–14.

Peteranderl S (2019) Wie finnische Häftlinge einen Algorithmus trainieren. *Der Spiegel*, 29 March. Available at: www.spiegel.de/netzwelt/web/finnland-wie-haeftlinge-einen-algorithmus-trainieren-a-1260249.html (accessed 7 September 2021).

Prison Policy Initiative (2020) Mass Incarceration: The Whole Pie 2020. Available at: www.prisonpolicy.org/reports/pie2020.html (accessed 7 September 2021).

Reisdorf BC and Jewkes Y (2016) (B)Locked Sites: Cases of Internet Use in Three British Prisons. *Information Communication and Society* 19(6): 771–86.

Rhodes LA (2004) *Total Confinement: Madness and Reason in the Maximum Security Prison.* Berkeley, CA: University of California Press.

Roberts ST (2019) *Behind the Screen: Content Moderation in the Shadows of Social Media.* New Haven, CT: Yale University Press.

Ruckenstein M and Teppo A (2005) *Vankien väliset valtasuhteet ja väkivallan pelko suljetussa vankilassa.* Helsinki: Rikosseuraamusvirasto.

Scholz T (ed) (2013) *Digital Labour: The Internet as Playground and Factory.* New York: Routledge.

Terranova T (2000) Free Labor: Producing Culture for the Digital Economy. *Social Text* 18(2): 33–58.

Vainu (2019) Vangit kehittämään suomalaista tekoälyä. Available at: www.vainu.com/fi/blogi/vangit-kehittamaan-suomalaista-tekoalya/ (accessed 7 September 2021).

Vallas S and Schor JB (2020) What Do Platforms do? Understanding the Gig Economy. *Annual Review of Sociology* 46: 273–94.

13

INVESTIGATING ADM IN SHARED MOBILITY

A design ethnographic approach

Vaike Fors, Meike Brodersen, Kaspar Raats, Sarah Pink and Rachel Charlotte Smith

The automated decision-making (ADM) systems that are invested in emerging transport technologies are designed to variously replace our actions when driving and to further enable us to combine and share different modes of transport. Recent discussions and debates concerning the ethics, sustainability and responsibility issues related to ADM have called for attention to the social implications and possible unexpected outcomes of its implementation in everyday life (AlgorithmWatch, 2019). However, algorithms for ADM-powered mobility solutions are rarely being developed with the social life of the end-users in mind, but rather in confined laboratory-like settings (Raats et al., 2020). For instance, our existing research has shown how in such lab studies algorithm developers put themselves into the role of the future users, to focus mainly on the momentary and individual use of the technology, with the objective of making it as efficient and easy to handle as possible.

In this chapter, we demonstrate how a design ethnographic approach to future algorithm-powered mobility solutions opens up possibilities to research social implications of ADM from a situational perspective, by investigating the context in which ADM is deployed rather than simply observing the technology itself and how it is used. We do so by contrasting everyday mobility decision-making (we call it EDM) with technological ADM solutions that have been developed for connected and shared transport solutions in an envisioned new 'mobility as a service' paradigm (Wong et al., 2020), to be able to discuss implications for further development of human-centred artificial intelligence (AI) in transport. The methodology and empirical insights described later derive from our project *Design Ethnographic Living Labs for Future Urban Mobility – A Human Approach* (AHA II). In this project, we combined ethnographic, co-design and Urban Living Lab approaches to engage communities and citizens in the design of future mobility services, based on local knowledge, community values and people's anticipations and expectations

DOI: 10.4324/9781003170884-17

about future smart mobility technologies. In this chapter, we concentrate on how we combined ethnographic fieldwork of people's EDM with future-oriented probing workshops to better understand the context and social implications of future ADM-powered solutions to shared and connected mobility.

The nearly universal use of smartphones has been hailed 'as the single greatest innovation for transportation in the last decade' (Wong et al., 2020: 1). Furthermore, the development of autonomous vehicles (AVs) and app-based applications for shared and connected mobility has been described as 'a new paradigm where mobility is no longer consumed as an asset (i.e., based on private vehicle ownership), but rather accessed on demand' (Wong et al., 2020: 1). Mobility, in this emerging paradigm, is developed as a service, where the user is expected to receive information, book and pay for a choice of different mobility services through an integrated digital platform, defined as Mobility as a Service (MaaS) (Mladenović, 2021). These ideas are fuelled by the growth of urban planning for city centres without privately owned cars and subsequently new infrastructures for combined modes of transportation that pull together a network of bike-sharing, scooters, buses, trams and automated and connected vehicles. From a technical perspective, shared and combined mobility systems for transporting people aim to minimise the number of vacant seats in vehicles in order to reduce the number of used vehicles, using concepts such as ridesharing, carpooling and car-sharing, managed by a growing number of on-demand app-based services (Curtis Lesh, 2013). At the heart of the emerging strand of research on efficient transport systems lies the development of algorithms for planning and operating such systems. Through simulations and data analysis, researchers hope to create options for people's travel that are so efficient and optimised that they will support the preferred choice of transport (Furuhata et al., 2013; Mourad et al., 2019). However, as we will demonstrate through our following examples, sharing and combined transport is far from solely being a technically driven practice since sharing practices are closely tied to the relational and social dimensions of the context in which it is embedded.

The AHA II project focuses on mobility within a mile from people's homes. Within urban planning and transportation research, this part of people's everyday mobility, as shown in Figure 13.1, has been pinpointed as a challenge for the transformation from privately owned cars to shared and connected mobility. From this technologically driven perspective, it is believed that poor connections from public transport nodes to people's homes are the main reason for people's preference for the privately owned automobile (Curtis Lesh, 2013; Shaheen and Nelson, 2016; Mohiuddin, 2021; Lu et al., 2021).

Current technological transport research and development focus on the possibilities of shared autonomous and connected vehicles to fill the alleged gaps in transportation systems during the first and last mile of travel (Gurumurthy et al., 2020; Ohnemus and Perl, 2016). The design ethnographic approach in the AHA II project moves beyond solely technology-driven optimising solutions by taking into account the experiences of people, families and community and including mobility practices that do not necessarily involve connectivity and data analytics. In

FIRST MILE   LAST MILE

FIGURE 13.1 The 'first mile and last mile challenge' refers to access and service quality at the outset of users' journeys. It refers to transport options used during the 'last mile' of urban commuting and is connected to debates on automated, connected and shared ADM-powered vehicles and services and what is needed for people to trust these.

Source: Designed by Kaspar Raats.

doing so, the project aimed to generate a locally grounded, in-depth understanding of travellers' practices, experiences and EDM. The ambition was to investigate the context of future ADM mobility technologies, in order to reveal any contradictions between the design vision imbued in the technologies and people's everyday mobilities. These tensions, between the design of technological systems and the ways people use them, make future mobility a good example of how ADM-powered automated and connected vehicles can be grounded in real-life situations that are not limited to automated and digital solutions. Thus, it provides opportunities to investigate the relationship between the technical design of ADM and social and real-life-based EDM, to subsequently create mobility solutions that resonate with local values and priorities.

A design ethnographic approach to ADM in everyday mobilities

Our interest in understanding how ADM-powered mobility would be adopted and appropriated among passengers in future shared and connected transport systems. This led us to develop a design ethnographic approach to investigate the context of ADM in everyday lives by combining the practical methods of ethnographic research into existing practices, routines and local knowledge and values with future-oriented co-design activities and probing.

Design ethnography is a methodology used across technology design, design anthropology and other participatory design disciplines in academia as well as in industry and consultancy contexts. As a practice, it can involve engaging ethnographic methods in order to understand everyday life circumstances and blending these with design methods, including design futures workshops, prototyping or speculative scenario creation. Design ethnography is often intended to be applied and interventional rather than simply a process of discovery and reporting. However, design ethnography is used differently across different disciplines because it is always made meaningful through the specific research questions, approaches to ethnography, analytical concepts and theoretical paradigms that shape research projects, findings and interventions. For example, when developed as part of design anthropology research (Smith and Otto, 2016; Pink et al., 2020), design ethnography is likely to take on board the critical perspectives of that subdiscipline, including critiquing and undermining narratives of technological solutionism through ethnographic attention to everyday experience and imaginaries and participatory design practice (Chapter 2, this volume).

This is the case in the AHA II project, where design ethnography brings together the theory, methods and intervention of ethnography and design to create a collaborative approach that involves both citizens and stakeholders from cities, public transport and the automobile industry. It involves not only using ethnographic methods, interviewing people and following them in their daily lives and communities but also working with participants and stakeholders in workshops, to co-create knowledge, imagine future technologies and codesign prototypes and services. The AHA II Urban Living Lab approach is closely related to human-centred and co-design approaches to cross-sector development, integrating research and innovation processes in real-life communities and settings (Marvin et al., 2018).

The AHA II approach as shown in Figure 13.2 entails a critical understanding of dominant existing and imagined future shared mobility systems powered and optimised by ADM-driven technologies, in order to re-frame what have become

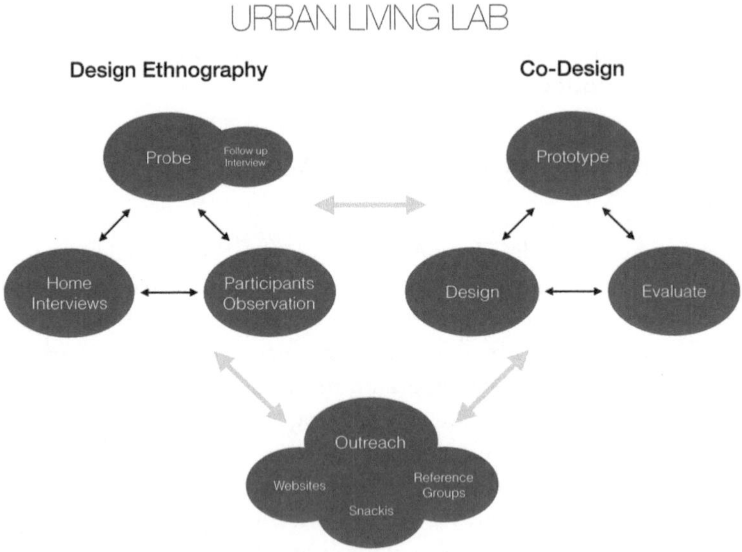

FIGURE 13.2   AHA II has developed a design ethnographic urban living lab approach to exploring future mobilities together with citizens in two residential areas in Gothenburg and Helsingborg in Sweden. The approach brings together a collection of methods and techniques to support human-centred activities and perspectives to innovation situated in a real-world context.

Source: Designed by Esbjörn Ebbesson.

FIGURE 13.3   One of our Urban Living Labs in the AHA II project, Bergum Gunnilse, a peri-urban area outside of Gothenburg.

Source: Map is produced under the Open Database License (ODbL) by OpenStreetMap Foundation (openstreetmap.org) and made available under the CC BY-SA license.

standard, one-size-fits-all solutions to products and services. In turn, these re-framings can be turned into shared mobility systems that attend to the needs revealed by our explorations of the context of ADM-technologies in people's everyday lives, their experiences, routines and foundations for habitual decision-making. In a context where shared and connected mobility systems are advanced in combination with AVs to address the 'first and last mile challenge', investigating shared mobility practices became a key strategy through which we grounded ADM research in concrete situations. Moreover, we situate questions about shared mobility in a specific *place*, showing how its specific material and social qualities influence future mobility and mapping out what happens within the space of the 'first and last mile'. In the following two sections we present our ethnographic fieldwork in a residential area in the outskirts of Gothenburg in Sweden, and our subsequent probing activities and workshops.

The socio-spatial dynamics of choosing modes of transport – complicating the first and last mile

Our fieldwork was undertaken in Bergum Gunnilse (see Figure 13.3), an area composed of a set of clusters of residential housing, in a hilly semi-rural landscape and stitched together along a main road that connects the areas and links to the city.

In a first ethnographic research stage of the project, we combined individual online interviews with an on-site visual ethnography that involved following the participants through their neighbourhood. A total of 20 participants were involved in this stage; they were recruited by the snowball method through local neigh-bourhood associations, as well as recruitment events outside the local supermar-ket. Aged 14–77 years, the majority of the participants were working parents of school-aged children, most of them lived in detached houses with one or several cars. In our online in-depth interviews, we focused on participants' biographical narratives, residential trajectories, perceptions of their neighbourhood and existing mobility and sharing practices that inform their everyday mobilities. Through these interviews, we learned about participants' EDM, their motivations and the social context for their choosing specific modes of mobility, as well as their representa-tions of different forms of (future) mobility.

We situated these elements within the specific surroundings that informed how participants envisaged mobilities through a method of what we call 'two-car drive-alongs'. This involved participants driving their own cars while two researchers followed them in a second car. The participants chose the starting point and guided the researchers through a selected area. They determined their routes in relation to relevant places and roads identified through a set of initial questions posed by the researchers. While driving, participants and researchers communicated via mobile phone and the whole encounter was both video and audio recorded. We used a Volvo XC90 hybrid as the 'following car', an iPhone 5 connected to the car's SPA [Scalable Product Architecture] infotainment system, using the car's microphone and speakers to interview participants. The interview was recorded using an audio recorder (Sony icd-ux570) placed at the centre of the car. The leading car and the space around were filmed with a GoPro Max 360 camera. We experimented both using a static camera positioned with a suction cup on the windshield attached behind the rear-view mirror and using an arm/tripod to film following partici-pants' indications from the passenger seat. This enabled us to address the particular local conditions and gave us specific insights into the processes of anticipation, negotiation and decision-making involved in navigating the local area, how its material features impact the organisation of mobility and the limits of car travel in this context.

Our research focused on a neighbourhood which is too dispersed to be walkable and where car travel is the dominant practice. Rather than attempting to reproduce 'naturally occurring' mobilities, the two-car-drive-along technique produced a situ-ation where participants are invited to identify and string together the places and routes most important to (their) mobility in their area. Guiding a second car in con-voy invites participants to make decisions about relevant places and questions and to make explicit self-evident practices and embodied knowledge about the place. This approach was tailored to provide meaningful interpretations of the layout of the spe-cific area and how these affect mobility decision-making. Interviews and drive-alongs allowed us to learn about: how participants' existing and imagined mobility decision-making practices were embedded in the socio-spatial context; how social relations and the specific qualities of local space intervened considerably in both their choices of modes of transport and in how they envisioned ADM. Indeed, while in dominant industry and policy narratives the 'first and last mile' often appears as an opportunity, rather than a challenge, our ethnographic research demonstrated that the first and last mile of people's travel was a dense and socially complex space in a way that moreover challenges the concept of the 'last mile challenge' in itself.

We argue that attention to EDM – in the context discussed here in the form of ongoing everyday mobility decision-making – is vital for understanding the context of future ADM implementation. In this section, we demonstrate ethnographically how such decision-making evolves *as part of* the social and physical environment as it dynamically unfolds when people go about solving their everyday logistics. Our ethnographic findings showed how the transportation decisions people make in the first and last mile of their daily travel are formed as part of a complex and dynamic web of socio-spatial relations. How people organise their first and last mile of travel to and from their homes is embedded in individual, social, institutional as well as physical contexts. For instance, the last mile may be intensely invested as a personal time-space that serves multiple purposes and/or marks transitions between social times. One of our participants, Amanda, uses the last stretch to work to walk and have time for herself, despite there being faster options available:

> I take the Blå Express to Svingeln and then I walk . . . it takes 25 minutes to walk to my work. So it's good because you get some exercise and you are really . . . prepared to start working when you arrive . . . I have music or a podcast in my ears and I walk pretty fast. So I go ten minutes earlier from home so I get this music time. Also often in my work it is also nice to close what has happened . . . and walk it off.
>
> *(Amanda, 41, 4 kids)*

However, the last mile may also become an opportunity for socialisation. As Simon mentions, 'when you live here, you will get to know people, everybody knows each other and people speak to each other in the street'. Thus, Simon coordinates his walking to the bus stop to be on the same bus as his friends and uses this time to socialise.

In many cases, the first and last mile question is made partly redundant by single-mode transport, especially car travel, for which, in many cases, the last mile 'challenge' is not the decisive factor. This is in part explained by the fact that usually combined trips become concentrated within the first and last mile space, requiring a certain level of coordination which extends beyond individual decision-making.

INTERVIEWER: okay and so when you go to school you take the bus as well, right?
SIMON: usually- when they used to drop the dog off at the.-
MOM: . . . the kindergarten for dogs
SIMON: . . . I usually go with them there and sometimes they drop me off at the bus so then I take the bus to school. . . .
MOM: if it is very late then we- we dropped him off at school; he is quite tired in the morning so it happens quite often that we take him [Simon] to school first and then drop off the dog

Approaching future mobilities through interviews also revealed how people's imagined 'needs' for future automated shared or on-demand mobility are inseparable from much wider social, spatial, institutional context. For instance, since Pernilla's children lost access to the school bus, rather than cycling as she used to, she combines her trip to work with driving them to school and adjusts the beginning of her workday accordingly.

> [My children are] 9, 11, 13. And before, the local authorities paid for a taxi . . . to school but this year they didn't get it. So if they are going by themselves . . . they have to wait 50 minutes at the school. So we usually drive them now, which means I have to drive or my husband had to drive, and then come to work later. . . . they really would like to take the bicycle I think, but it's not possible with this traffic. It's a lot of traffic and there is no bicycle path, and it's quite a dangerous road.

Pernilla's family's experience also highlights how choices concerning modes of transport are far from being a matter of individual choice or personalisation but are embedded in complex family logistics, spatialities and diverse social relations. In the area, coordinating and facilitating children's mobility is a central motivation for multiple car ownership within households. Antonia drives her kids to school and drops off their bikes on the way down at the bus stop on days where they finish early so they can take a tram and bus back and cycle the last 3 km from the bus stop home along a dirt road. In the winter, she prefers to coordinate with her husband to pick up the kids either at school or at the bus stop.

> We always have the bike stand in the back of the car, because I always need to drive the bikes very often. But it's less now during winter or fall, because when they come home it's dark, this a dirt road – so there are no like lights or anything – and forest.
>
> *(Antonia, 39, 2 kids)*

Moreover, the specificities of the topography of first and last miles critically impact on how travel is envisioned and organised. In Bergum Gunnilse, the first mile would typically be the distance home from the bus stop on the main road, which could involve a 5 km uphill hike in low visibility without a pedestrian path. Steep hills, narrow roads that struggle to absorb the population growth in the area, darkness, weather and wilderness were frequently mentioned to justify individual car use. Moreover, the lived environment was also part of participants' mobility and was a consideration when they discussed how they envisaged future everyday local uses of technologies like AVs.

> I just have a hard time seeing how self driving cars would work in real life. I would want to know the technology behind how it works if unpredictable things happen around the car. And if you would go on a tiny road, like the

last two kilometres to the Lake where I like to go . . . it's looking out for animals, since it's in the forest. And then also driving up the steep hills with the tiny stones in the ground I need to make sure that I can drive up safely, and not having the car getting out of my control and sliding down the hill again.

(Emma, 20)

The combined and complexly coordinated mobilities, including in the first and last sections of journeys, and local mobilities, demonstrate how the framing as a 'first and last mile challenge', implying singular, point-to-point mobilities, is in itself problematic.

Probing future ADM-powered mobility solutions – complicating sharing

While elements of EDM can be prompted in interviews and on site visual ethnography techniques, experiences of speculative future ADM-powered mobility solutions are more aptly investigated through probing techniques.

Sharing is promoted in dominant narratives in the form of the emergence of a 'sharing economy' (Pouri and Hilty, 2021) which is viewed as a 'solution' towards access and sustainability through the application of ADM-powered digital services (Wong et al., 2020). However, our interview research revealed that sharing is already practised in a variety of EDM forms, most of which are quite different from the 'sharing economy' understanding of commercial transactions monetising underused assets. Sharing is mostly limited to relevant groups and communities, where it is imbued with symbolic meaning and serves a function of social integration. To take this further, we developed probing techniques to explore how sharing might be envisaged as part of everyday mobility futures. Probing (Gaver et al., 1999) does not only question experiences, representations and expectations regarding ADM and sharing practices but also proves more effective in eliciting future imaginations (Hutchinson et al., 2003) while allowing people to think about their experiences in a new way and thereby propose improvements (Mattelmäki, 2006; Wallace et al., 2013).

To be able to reflect our ethnographic insights in people's everyday transport decision-making in more technology-driven design visions of efficient and optimised ADM-powered mobility solutions, we combined the ethnographic interviews with participatory co-design online workshops to probe towards future imaginaries of relevant mobilities. Participants in the ethnographic fieldwork were invited to recruit friends, neighbours, colleagues or teams members that shared activities or residential spaces in the area to participate in workshops where we first talked about their common experiences in the area and past and present sharing practices, then introduced the idea of a shared autonomous pod as a backdrop to co-design ideal future mobility solutions. We conducted seven workshops with an average of three participants in each.

We chose to structure the probing around a shared autonomous pod since it resonates with future visions of self-driving vehicles that bring people to or from other modes of transportation in technology-driven imaginaries of futures with no privately owned cars, as well as with our ethnographic insights in how people in the area coordinated shared transport in everyday mobilities. In this way, we could probe how a future ADM-powered technology could be integrated in existing mobility practices. The activity was supported by a map of the area inserted into an online collaboration platform Mural (see Figures 13.4 and 13.5) that allowed participants to insert drawings and Post-it notes and thereby superimpose layers of present and future mobilities on the geography of the area; ideas were collected and readjusted on the same platform throughout the discussion. Basing the activity on existing relationships and shared experience further reduced abstraction and grounded imaginary situations in real-life social contexts. As shown in Figure 13.4, we asked the participants to visualise destinations they visited, areas they had mentioned to be challenging and other places they felt relevant. To connect these to future visions we asked them to also, for example, mark potential self-driving vehicle pickup spots, destinations the vehicles would enable to visit and places where they thought the self-driving vehicles would struggle.

Bringing together participants who already share transport in our co-design probing workshops suggested that some of the existing sharing practices and the symbolic and social meanings they involved (revealed in our ethnographic work discussed earlier) might be supported by ADM. However, the use of automated technologies and platforms to generate shared mobility systems and practices was restricted by questions of participants' trust in ADM to be able to monitor the social dimensions of travel and concerns about other humans. A key example of

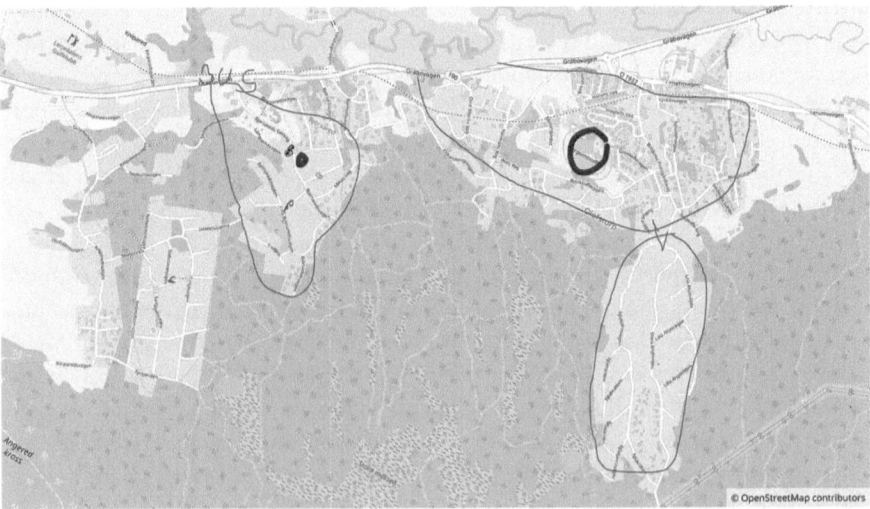

FIGURE 13.4 Map drawings from a workshop with Felix, Jonas and Olaf, whose children play in the same football team.

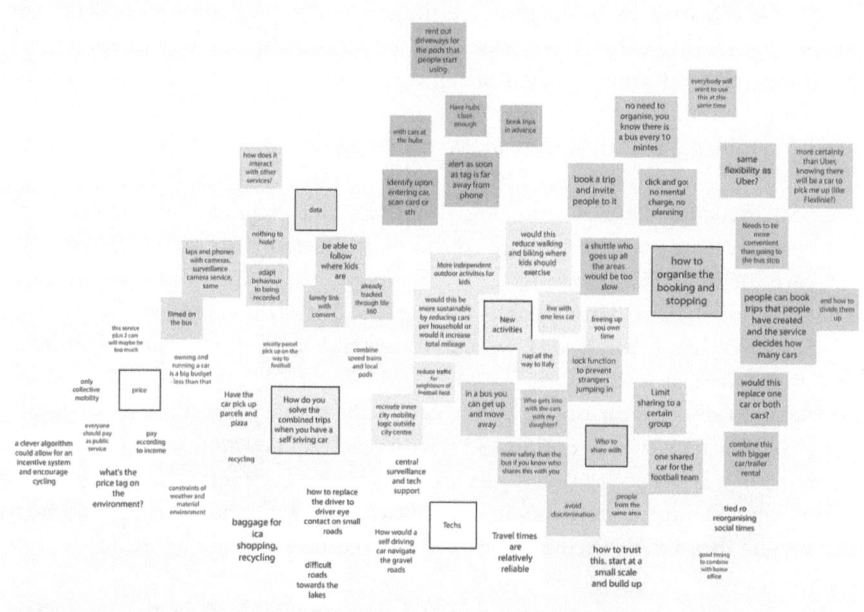

FIGURE 13.5   Idea collection and themes from workshop with Lina, Yvonne and Elsa. Ideas were collected by the researchers throughout the discussion and subsequently grouped into themes, edited and completed by the participants.

Source: Reproduced here with the permission of the workshop participants.

this is the shared responsibility of driving children to sports practice, which occupies significant portions of parents' time in the area.

FELIX: One case would be to go to the football practice; we are all involved in that. Already today, Olaf and me, we live in the same area so we usually drive together and depending on timings we have some kind of sms group where we just call out and ask if somebody wants to come to the training. It would be a pretty straightforward case to just transform that to a self-driving service . . .

JONAS: How do I know that the car is not letting anyone into the car if I send my kids to their practice? How do I know if the kids arrived to the practice?

FELIX: Yeah, how do I know about the human security? Do I know if it's a private drive, so no one else hops on? If it's Jonas' kids, then it's ok.

OLAF: The pod could only open the doors at certain geographical locations. . . . Then you pre-program the pod to get a higher security. Then the car could go on a 'milk-run', to just go around and pick up people.

FELIX: We have a football list, so we have names, and can we get a proposal of what kids will go, then the service could plan out the rides of where and who to collect. Then it would be a logistic support.

Sharing and trust in automation are thus mutually restrictive. In real-life contexts where sharing is involved, trust in other users is the primary theme that the discussions centred around when it came to using an AV service. In co-design workshops, when faced with concrete scenarios, trust in other humans tended to be more problematised than trust in AV technology.

YVONNE: Self-driving cars seem less dangerous than regular cars, or taking the bike. If in an accident you are still more protected than on a bike since you have metal shielding you.

LINA: It's not the accidents people are scared of, it's the people sharing the car. Scared of being harassed. I wouldn't be afraid sending my kids off in it, not around here at least.

YVONNE: Maybe the car can be connected to the activity so it only picks up people at the activity at a set time. So it won't pick up any adults. – This is booked just for this event, or this is just a public round for everyone. People should need to identify themselves in some way. . . . The regular one would be for anyone. But for recurring events, a dedicated round could be made.

Trust in AVs appears as processual and experience-based in the participant's stories. Rather than questioning the inherent qualities of the algorithm steering the ADM technology or the legitimacy and efficacy of the organisation developing it, participants stressed that the key condition to be comfortable with ADM technology is that they would have to 'see it at work' and progressively get used to it through repeated use in real-life situations.

JONAS: I should probably use this service quite a lot of times before I leave my kid to use the pod themselves. But my real need is to send them off by themselves, so I can stay at home and do other work, or drive one of my other kids. I would need to go with them the first time they use it.

These elements are reinforced by the idea that trust in ADM-powered AV services is often mediated and placed by one person for another person (a child, elderly parent, etc).

DAN: I've got a son, he goes to school in Gunnilse and then next year he is starting at the [school in the city]. We were kind of thinking of . . . getting him on the bus by himself. . . . So we just try to decide where's the fine line whether they are too young or not too young. The other thing is that, unlike his sister he is probably less focused in what he is doing and probably walking along holding his phone or something. Not having an eye on where the trams and buses are in town. . . . [W]hen she was on the tram she [his sister] did it really well, and the busses and everything. She is now quicker than the Resplanare [planning app]. . . . That is where we want to get to anyway with [him], but we have started to see that. yeah. we have to drive . . .

NINA: We also . . . our oldest goes to another school in Hisingen. She wants to take the bus [which] I think it's really good as she knows her way around Gothenburg now or this area. But I have the same feeling about our son who is in fourth grade. He would lose himself somewhere.

The relevance of ADM is also relational to how existing mobility patterns and sharing practices participate in social integration. For instance, when parents drive children to after-school activities, the shared mobilities involved are inextricable from their social groups, symbolic meanings and performative functions. 'Sharing' in this sense differs from the 'sharing economy' understanding of commercial transactions, as demonstrated in the following example:

AMANDA: In another group, some of the parents always took the best time and then we had to talk about it because nobody wants to go at 10 in the evening every week. . . . In this group, everybody says 'I can drive, I can drive' so we have . . . more the problem that people feel sorry 'Oh I haven't driven this week'.

LENA: I think [for us] it's the same as in Amanda's group. . . . I think it works fine but you need to take some time during the day to send texts and sometimes you don't know if anyone can take them or not like half an hour before they have to leave. So you need to text and organise. . . . they can always go by bus so if no one can drive. And sometimes I think 'why don't they take the bus all the time' – because they can do that. I think we are so involved in our kids and we really want to be part of it and we really want to show each other that we are good parents and I think that is part of it. I think it would be good if someone just said 'they can go by bus, that's fine'.

Our design ethnographic approach demonstrates how the realistic possibilities for future shared ADM-powered mobility solutions are best envisioned in the context of a wider set of social and socio-spatial relations and circumstances. This means that through our focus on understanding everyday transport sharing, we realised that successful ADM implementation has to be guided by how trust in other users develops in real-life situations and the fact that people need to 'see it at work' and progressively get used to it through repeated use in real-life situations.

Re-framing ADM-powered mobility

The design ethnographic research in the AHA II project shows that optimising the first and last mile by creating seamless efficiency between different modes of transportation is not a clear-cut answer to existing problems and needs, because such an approach extracts expectations and representations from the concrete socio-spatial situations and questions in which they occur. In Bergum Gunnilse, the last mile is a challenging trip through difficult terrain that discourages light mobility or transit use. Given the geography, the last mile may require considerable effort.

Most importantly, the first mile is a matter of coordination and family logistics. This often results in choosing the car which will be used as part of a strategy to combine trips. Moreover, the last mile is not always a problem that needs solving but rather an opportunity for sociability, exercise or quality alone-time – it involves central features such as popular shortcuts and informal meeting places. Within the last mile, people find close neighbourhood ties, or a pre-school, playground or a bus stop.

Taking this physical and social context into consideration suggests a different framing for ADM to that suggested by the technology-driven agendas behind algorithm development, which see it as being designed to serve individualised, seamless and momentary uses (Raats et al., 2020). Our design ethnographic approach points out a series of contrasts between development and user rationalities between ADM and EDM. Where developers focus on the inherent qualities of algorithms rather than user needs and real-life user contexts, people focus on how the algorithms perform in their context. To participants, potential use was more dependent on the ways in which they could modulate encounters with other users (of the AV and public road space) in already existing mobility decision-making practices, than it was on the quality of the automation itself.

We suggest that this is where a design ethnographic approach that combines ethnographic research with future-oriented probing can become useful in the development process of future ADM mobility technologies. In turn, this approach can be developed in response to recent calls for human-centred algorithm development (Baumer, 2017). Ethnographic interviews show that sharing practices (digitally supported or not) exist and are meaningful. However, probe workshops show the possibilities and limits of extending and automating such sharing practices in the future, since sharing is mostly limited to relevant groups and communities, where it is imbued with symbolic meaning and is part of social integration. A viable combination of technology development and design ethnography could be to create iterations of what is known of existing situated practices that produce use cases for developers to process and then deliver ideas for probing workshops.

Based on what we have learned through our studies of mobility algorithm developers' individualistic framing of the perceived user, tested in confined spaces to reduce the level of complexity, it is clear that the outcomes of implementing such algorithms into the socially and materially embedded first and last mile are uncertain. Our research demonstrates that a technology that works fine in the confined spaces of algorithm development is by no means guaranteed to solve any problems in everyday social life. If the problems ADM is set to solve in terms of optimisation and efficiency are not anchored in how people would activate it in their daily routines, it will only create new problems.

Acknowledgements

This chapter presents results from the project *Design Ethnographic Living Labs for Future Urban Mobility – A Human Approach* (AHA II), which is a Swedish collaborative project between Halmstad University, Volvo Cars, City of Gothenburg, City of Helsingborg and the public transport companies Skånetrafiken and Västtrafik. Project AHA II is part of Drive Sweden's strategic project portfolio, funded through VINNOVA between 2019 and 2022. Drive Sweden is one of the Swedish government's 17 Strategic Innovation Programs (SIPs).

References

AlgorithmWatch (2019) Automating Society – Taking Stock of Automated Decision-Making in the EU. Available at: https://algorithmwatch.org/en/automating-society-2019/ (accessed 28 May 2021).

Baumer EP (2017) Toward Human-Centered Algorithm Design. *Big Data & Society* 4(2). https://doi.org/10.1177/2053951717718854

Curtis Lesh M (2013) Innovative Concepts in First-Last Mile Connections to Public Transportation. In: *Third International Conference on Urban Public Transportation Systems*. Paris, 17–20 November. https://doi.org/10.1061/9780784413210.007

Furuhata M, Dessouky M, Ordóñez F, Brunet ME, Wang X and Koenig S (2013) Ridesharing: the State-of-the-Art and Future Directions. *Transport Research Part B* 57: 28–46. https://doi.org/ 10.1016/j.trb.2013.08.012

Gaver B, Dunne T and Pacenti E (1999) Design: Cultural probes. *Interactions* 6(1): 21–9. https://doi.org/10/cfnt8c

Gurumurthy KM, Kockelman KM and Zuniga-Garcia N (2020) First-Mile-Last-Mile Collector-Distributor System Using Shared Autonomous Mobility. *Transportation Research Record* 2674(10): 638–47. https://doi.org/10.1177/0361198120936267

Hutchinson H, Mackay W, Westerlund B, Bederson BB, Druin A, Plaisant C, Beaudouin-Lafon M, ConverSy S, Evans H, HaNsen H, Roussel N, EiderBäck B, LindquiSt S and Sundblad Y (2003) Technology Probes: Inspiring Design for and with Families. In: *CHI'03 Proceedings of the SIGCHI Conference on Human Factors in Computing Systems*. Fort Lauderdale, FL, 5–10 April, 17–24.

Lu Y, Prato CG and Corcoran J (2021) Disentangling the Behavioural Side of the First and Last Mile Problem : The Role of Modality Style and the Built Environment. *Journal of Transport Geography* 91. https://doi.org/10.1016/j.jtrangeo.2020.102936

Marvin S, BuLKeley H, Mai L, McCormick K and Voytenko Palgan Y (2018) *Urban Living Labs. Experimenting With City Futures*. London: Routledge.

Mattelmäki T (2006) *Design Probes*. Publication Series of the University of Art and Design Helsinki A, 69/2006. Aalto: Aalto University.

Mladenović MN (2021) Mobility as a Service. In: Vickerman R (ed) *International Encyclopedia of Transportation*. Amsterdam: Elsevier, 12–18.

Mohiuddin H (2021) Planning for the First and Last Mile: A Review of Practices at Selected Transit Agencies in the United States. *Sustainability* 13: 2222. https://doi.org/10.3390/su13042222

Mourad A, Puchinger J and Chu C (2019) A Survey of Models and Algorithms for Optimizing Shared Mobility. *Transportation Research Part B* 123: 323–46. https://doi.org/10.1016/j.trb.2019.02.003

Ohnemus M and Perl A (2016) Shared Autonomous Vehicles: Catalyst of New Mobility for the Last Mile? *Built Environment* 42(4): 589–602. https://doi.org/10.2148/benv.42.4.589

Pink S, Osz K, RaaTs K, Lindgren T and Fors V (2020) Design Anthropology for Emerging Technologies: Trust and Sharing in Autonomous Driving futures. *Design Studies* 69. https://doi.org/10.1016/j.destud.2020.04.002

Pouri MJ and Hilty LM (2021) The Digital Sharing Economy: A Confluence of Technical and Social Sharing. *Environmental Innovation and Societal Transitions* 38: 127–39. https://doi.org/10.1016/j.eist.2020.12.00

Raats K, Fors V and Pink S (2020) Trusting Autonomous Vehicles: An Interdisciplinary Approach. *Transportation Research Interdisciplinary Perspectives* 7(c): 10. http://dx.doi.org/10.1016/j.trip.2020.100201

Shaheen S and Nelson C (2016) Mobility and the Sharing Economy: Potential to Facilitate the First- and Last-Mile Public Transit Connections. *Built Environment* 42(4): 573–88.

Smith RC and Otto T (2016) *Cultures of the Future: Emergence and Intervention in Design Anthropology*. London: Routledge.

Wallace J, McCarthy J, Wright PC and Olivier P (2013) Making Design Probes Work. In: *Proceedings of the SIGCHI Conference on Human Factors in Computing Systems – CHI'13*, Paris, 27 April–2 May. https://doi.org/10/ggc5sd

Wong YZ, Hensher DA and Mulley C (2020) Mobility as a Service (MaaS): Charting a Future Context. *Transport Research Part A* 131: 5–19. https://doi.org/10.1016/j.tra.2019.09.030.

14

AD ACCOUNTABILITY ONLINE

A methodological approach

Mark Andrejevic, Robbie Fordyce, Luzhou Li and Verity Trott in collaboration with Dan Angus and Xue Ying (Jane) Tan

Introduction

Advertising is undergoing a dramatic shift from public to private thanks to the prevalence of targeted, online commercial messaging delivered on personal devices. Whereas once upon a time, ads were widely publicly available – hence the association with 'publicity' and 'public relations' – when they migrate online ads go 'dark': they are only visible to those to whom they are directly targeted. This changes our everyday experience of advertising and has important consequences for long-standing social concerns about the potential pathologies of commercial messaging, including, notably, discrimination, stereotyping, predatory advertising, and the circulation of false and misleading information. Moreover, the lack of accountability of online advertising provides cover to advertisers to engage in activities that they would likely avoid if they knew they were subject to public scrutiny. Much of the research on so-called 'dark ads' has focused on political advertising (Saunders, 2020), but the impact of advertising reaches far beyond politics to reflect and reinforce a range of social values and associations (Ewen, 1974; Schudson, 1984). Thus, a shift in advertising techniques, practices and content has consequences that range beyond the realm of politics to society in general. This chapter explores possible methods for providing greater accountability for online advertising and thus for raising awareness of and devising responses to the dramatic shift in the advertising ecosystem resulting from the shift to online ad targeting.

Advertising as a cultural system

Despite the magnitude of the shift in contemporary advertising infrastructures and practices, there has been little systematic research on how historical concerns about the social issues associated with advertising are impacted by the changing ad environment.

DOI: 10.4324/9781003170884-18

This lack results in no small part from a structural change in the available archive. Thanks to the advent of technologies such as video and audio tape, as well as storage media such as microfilm and microfiche, there are enduring archives of advertising from the recent mass media era that can be drawn upon by researchers, media activists and the general public. When ads go online, however, they become both more copious in number and more ephemeral. One ad campaign may go through thousands or tens of thousands of variations, some of which are only delivered once and most of which are publicly inaccessible once they have finished their run. Because online ads are programmatic – that is targeted to particular viewers regardless of content – they cannot be retrieved by going back and viewing the content, as would be the case, for example, with print ads. A different person viewing the content, or even the same person going back to view the content, will likely see a different ad.

The result is that we are losing collective visibility into the advertising environment, and this makes it difficult to address the potential social issues it raises. We know, for example, that the long history of racism and sexism in advertising has helped reinforce the attitudes, associations and prejudices that support and enable violent and discriminatory policies and actions. In other words, the political impact of advertising is not limited to overtly political and issue-oriented ads. Racist, sexist and homophobic advertising, for example, all have a significant cultural role to play in the reproduction of structural forms of discrimination and the violence visited upon victimised groups. Australian Indigenous scholar Kathleen Jackson, for example, notes the connection between racist ads and harmful social policy in her discussion of the notorious Nulla-Nulla soap ad which personified 'dirt' in the form of an Aboriginal woman being beaten on the head under the brand slogan, 'knocks dirt on the head'. As she puts it,

> Advertisements, such as Nulla-Nulla soap, provided subliminal support to the colonial campaign to enforce European cultural and economic values. . . . A single complaint about the cleanliness of an Aboriginal child could result in the exclusion of Aboriginal children from school. This exclusion could establish neglect and allow . . . the removal of Indigenous children from their families.
>
> *(Jackson and Barnes, 2015: 73)*

Degrading images and dehumanising stereotypes go hand in hand with violent acts. In the realm of advertising, these messages are not framed in terms of politics or social issues, rather they form the taken-for-granted cultural background for everyday commerce. They enact a kind of mundane, quotidian, quasi-invisible ideology. In this respect, commercial – as opposed to explicitly political – advertising plays a central role in the everyday production of taken-for-granted assumptions. The cultural images a society feeds to itself through its commercial system do much more than sell products: they reflect and reinforce dominant social values and associations.

The challenge posed by the digital media environment is that it becomes more difficult to obtain public visibility for the commercial messages bombarding users on their personal devices. From the perspective of the individual user, there is exposure to more advertising and branding messaging than ever before (Carr,

2021). From the broader societal perspective, however, a fast-growing portion of these ads is effectively invisible. This invisibility has consequences for individual users as well – they may see the ads that have been targeted to them, but they have no way of knowing whether others are seeing the same ads – and if so, which others. A man who sees a job ad, for example, may not know that this ad has been targeted exclusively to men of a particular age. Someone seeing an ad for easy credit is unable to know whether the interest rate they are being offered is the same as that offered in ads to others. This invisibility is facilitated by the fact that we have typically treated ads as ephemera – we may want to keep a magazine article or record a TV show, but only rarely do we want to keep the ad or record the commercial. Ads are free riders on content and are treated as such: a necessary inconvenience that may on occasion provide some information of interest. When ads come and go in our online news feeds or in the interstices of the content we access online, we may pay little attention – but this does not mean that they are irrelevant forms of messaging. Quotidian ideology functions most effectively when it operates in the form of a background set of taken-for-granted associations (see, e.g. Billig, 1995).

Moreover, advertising has an important role to play as a form of shared commercial glue holding together the edifice of social media communication. As media scholars have long argued, commercial messaging played a central role in the rise of mass consumer society in the 20th century. The historian Jackson Lears, for example, contends that advertising has collaborated with other social institutions to promote, 'dominant aspirations, anxieties, even notions of personal identity' (1995: 2). In addition to its role in mobilising consumption to keep pace with the productivity of industrialised mass production, advertising has a broader cultural significance. As Michael Schudson argues, advertising 'may shape our sense of values even under conditions where it does not greatly corrupt our buying habits' (1984: 23). Specifically, he argues,

> Advertising, whether or not it sells cars or chocolate, surrounds us and enters into us, so that when we speak we may speak in or with reference to the language of advertising and when we see we may see through schemata that advertising has made salient for us.
>
> *(Schudson, 1984: 210)*

Given its central cultural role in shaping attention and reinforcing social trends, much work has been done on the role played by advertising in reproducing stereotypes, preconceptions and dominant meanings and associations. Scholars and researchers have explored the role played by advertising in shaping attitudes toward female body image and beauty (Kilbourne, 1990), racial preconceptions and prejudices (Wilson and Guttierez, 1995) and class (Marchand, 1985), among other areas of social life. As ads come to permeate contemporary life, the values and attitudes they select and reinforce become a core component of the informational atmosphere through which we move, in combination with the influence of the family, schools and other arenas of meaning-making and cultural production.

If advertising makes up a significant portion of the information to which we are exposed each day – one which has significant consequences in terms of directing attention and highlighting social values and the limits of representation – it needs to be available for public examination. This is the premise of the research methods described in this chapter, which are designed to increase public awareness about the online ad environment and to foster public discussion about how it is being used and abused. Advertising accountability tools address a structural change in the cultural environment. Much public attention has been devoted to the epochal change in how news and other forms of content are distributed – but given the historical role played by advertising in reinforcing cultural values and stereotypes, similar attention needs to be paid to advertising. Societies need to be able to reflect back on themselves in order to evaluate and assess the shared sets of values and concerns reproduced within their commercial media environment – an environment obscured by the rise of targeted advertising on personal devices.

Shifts in the everyday experience of advertising

In the mass media era, ads formed a part of a society's shared media culture. Because of the costs of production and the nature of the medium, the same ads were repeated over and over again. Anyone who grew up during the heyday of the commercial mass media can recall the jingles and images of the ads or their youth. These ads were openly available to public scrutiny – and for inclusion in public archives, which allowed them to serve as both barometers and reminders of shifts in cultural values and associations. A look through the advertising archive provides striking examples of these shifts. What were once commonplace representations of gender roles and relations, for example, may look appallingly sexist by contemporary standards. Some ethnic stereotypes that were once acceptable to the dominant culture no longer have a place in the public advertising landscape – as suggested, for example, by the decision of the Mars Inc. food company to drop the brand emblem of 'Uncle Ben' – a racist stereotype – from the packaging and branding of what used to be called 'Uncle Ben's Rice' (Booker, 2020). This is not to say that sexism and racism have disappeared from the contemporary ad environment but rather that the bounds of what is considered acceptable for mainstream consumption and mass marketing have shifted. There may well be some consumers who would be comfortable with the persistence of degrading stereotypes (of others), but mass advertising takes into consideration a general audience – and this is reinforced by the scrutiny of journalists, media watchdogs and public interest groups.

In the 1980s, for example, consumer advocacy groups expressed concern that tobacco and alcohol companies were deliberately targeting billboard ads to low-income, predominantly African American neighbourhoods. These ads were easy to see by moving from one neighbourhood to another because they were publicly displayed. The disingenuous response of industry representatives was to turn the accusation back on their critics by accusing them of racism for implying that minority populations were more susceptible to advertising appeals. The city of St. Louis, for example, conducted a public inventory of its billboards that discovered not only that billboards were more prevalent in Black neighbourhoods but also that the percentage of ads devoted to cigarettes and alcohol in Black neighbourhoods was almost twice that of white neighbourhoods (The Media Business, 1989). It was relatively easy to conduct such a survey because the billboards could be viewed and counted by anyone in their vicinity.

The rise of targeted advertising delivered to personal devices via digital media inaugurates an era of mass customised messaging that increases the variety and turnover of ad campaigns. The fact that it is relatively inexpensive to automatically create and test hundreds or thousands of ad variations almost instantaneously results in dramatically higher turnover rates. The fact that people are spending more time online, then, means they can be exposed to a greater number and variety of ads than ever before – fragmenting what might be described as the formerly shared culture of public advertising. Moreover, the need to cater to shifts in dominant sensibilities is obviated: if advertisers know that certain audiences will be comfortable, for example, with stereotypes and associations that might trouble, alienate or anger others, they can adjust targeting strategies accordingly to ensure that ads are delivered only to those unlikely to protest. The affordances of customisation exacerbate informational asymmetries in the advertising environment: platforms (and, in some cases, advertisers) know more than ever before about consumers, who, in turn, know less than before about why they are being targeted and what ads others are seeing. High turnover of ephemeral ads makes it harder than ever before to create a shared public portrait of the advertising environment.

It is very difficult to tell, for example, whether advertisers are resuscitating racist appeals for particular audiences – or creating new ones – because we have such limited visibility into online advertising. It took an investigation by independent journalists attempting to buy different types of ads to discover that Facebook made it possible to deliver job ads that discriminated by age – in violation of federal law (Angwin et al., 2017). Facebook eventually paid a $5 million fine and agreed to create a separate category of ads for jobs, housing and credit that prevented targeting by protected categories (Kaya Yurieff, 2021). However, even this solution does not necessarily eliminate discrimination in advertising, given that Facebook's algorithms have been demonstrated to customise ad delivery based on information about past response patterns. If, for example, a previous ad, similar in content or appearance, was primarily clicked on by men, the algorithm would target the new ad primarily to men, even if the advertiser had not requested it to do so (Ali et al., 2019). In this way, algorithms might perpetuate historical forms of discrimination on their own – without any intentionality on the part of advertisers.

Tools for ad accountability

The lesson of such examples is that it is not enough to protect against the intent to discriminate, nor is it enough to trust that commercial platforms will follow the law or their publicly stated commitments. Even after Facebook claimed to have changed its ways, the investigative journalists at *ProPublica* were able to purchase housing ads that 'specifically excluded "African Americans, mothers of high school kids, people interested in wheelchair ramps, Jews, expats from Argentina and Spanish speakers"' (Larson, 2017). Many of these forms of discrimination are illegal, insofar as they involve protected categories of ads (housing and job) and persons (ethnicity and disability). But even if Facebook were to address these, it is legal – if potentially socially detrimental – to discriminate in the delivery of non-protected categories of ads and persons. It would not be illegal, in many jurisdictions, for example, for Facebook to disproportionately target alcohol advertising on the basis of race. Nor would it be illegal for it to engage in the forms of stereotyping in ad content that have become socially objectionable when publicly displayed. In the interest of niche marketing, the mobilisation of offensive stereotypes might even be treated as an effective strategy for appealing to overt racists who pride themselves on their 'political incorrectness'. The point to be made here is that the online setting bypasses the forms of social accountability whereby we decide what, as a society, we consider to be acceptable standards for treating one another.

When it comes to these forms of online opacity, a range of strategies have been developed to provide accountability and transparency where areas of social concern are at stake. These might be categorised in terms of whether they focus on inputs, processes or algorithms. In the former case, when the data that are used to train automated systems have been found to contribute to biased outcomes, the goal has been to create ethically sourced and diverse datasets that are screened for bias – and to ensure that the teams creating and tuning the systems are themselves diverse (*Appen*, 2021). Data is only part of the equation: providing accountability and transparency also means securing some form of access to the processing system. When it comes to algorithmic decision-making systems, for example, there has been a push toward 'explainable' AI (Goebel et al., 2018). In many cases, these systems are commercial and proprietary, which means they are unavailable for public inspection. Google's search algorithm, for example, lies at the heart of the company's commercial property and thus its value. In this respect, transparency and the accountability it affords can be difficult to achieve. Similarly, some processes can be protected by encryption – such as communications or transactions that take place on the so-called 'dark web'. In such cases, governments have, for example, proposed requiring 'back doors' that allow access to encryption, spawning an ongoing debate about security.

When it comes to automated sorting and decision-making processes, one way to address the 'black boxing' of automated systems – whether this is due to commercial restrictions or to the sheer complexity of the operations involved – has been to audit their outcomes (Sandvig et al., 2014). The advantage of auditing is that it focuses on the core area of concern when it comes to addressing the opacity of automated systems: their results. Databases that have been scoured and checked for diversity and inclusion can yield flawed or detrimental results as can transparent algorithms. Auditing zooms in directly on the results and can thus point back at problems in the data collection and processing stages. The drawback of auditing is that it can be difficult to achieve at scale. Auditing all of Trump's Facebook ads during the 2016 campaign, for example, would be a daunting task, given that his digital media consultant, Brad Parscale, claimed the campaign served up to 30,000 different, unique, ads *daily* in the weeks directly preceding the election (Green and Issenberg, 2017). However, even a partial glimpse at outcomes is better than none. This has been the philosophy of a range of accountability organisations such as AlgorithmWatch in Germany, ProPublica and the NYU Ad Observatory in the USA, all of which rely on 'data donation' tools which allow members of the public to share information with researchers so as to provide accountability for the operation of targeting algorithms.

The Facebook ad collector

It is perhaps not surprising that several data donation projects focus on advertising, sponsored messaging or results for pre-scripted search engine queries. The challenge faced by researchers is how to provide accountability without being overly invasive. With this consideration in mind, the following sections focus on one methodology for subjecting commercial institutions to some form of accountability. This method builds on the *ProPublica* ad collection tool – which has also been used, in updated form, by *The Guardian* newspaper and by New York University's Ad Observatory – but adds a demographic component. This component addresses the fact that while the *ProPublica* ad collection tool (which also collected ads on Facebook) revealed the types of messages circulating online, it provided little insight into the patterns of demographic discrimination that characterised the circulation process. It did collect some basic information gleaned from the 'why are you seeing this ad' caption provided by Facebook, but this is selective and limited. It is unlikely, for example, that Facebook notified recipients of the Trump campaign's voter suppression ads that they were being targeted because they were African Americans in swing states (Green and Issenberg, 2016).

The goal of ad accountability methods is, in a sense, to 'humanise' the automated curation process. Given the mass scale of targeted advertising on platforms like Facebook, which reaches almost half of the world's population over the age of 13, all we can absorb as human auditors is a tiny subsection of the processes at work (Statista, 2021). Nevertheless, it is not impossible to envision the prospect of scaling up the accountability process with the assistance of automated forms of data analytics: using algorithms to provide insight into algorithms.

This approach to advertising accountability builds on the *ProPublica* tool, which is available in open source form online. We employed a software designer to redevelop a version of the tool that would operate as a plugin for the Chrome browser on desktop computers and which would gather only those ads that appeared in the feed of the default main Facebook page. This allowed participants to have some control over our data gathering process insofar as they could elect to uninstall the tool or use alternative browsers if they wanted to control their degree of participation in the research project. The tool relies on the fact that sponsored content must be indicated as such on Facebook – this means that we can program the tool to search the HTML code and determine which content has been labelled as sponsored. It is worth noting that Facebook does not encourage the use of the *Pro-Publica* tool and has repeatedly changed the way it codes the 'sponsored' tag in order to hide it from automated ad collection. We had to update the tool several times over the course of the year in order to adjust to changes by Facebook apparently designed to thwart accountability tools like *ProPublica*'s. The tool is not a mass data-scraping technology and thus, arguably, does not violate Facebook's stated terms of service. Rather it provides a way for individual users to decide to share information they receive online with researchers. In this respect, it replicates one of the main functions of Facebook: it is, at base, a tool for sharing content.

The main difference between *ProPublica*'s tool and ours is that we designed the installation procedure to collect voluntarily supplied demographic information about participants. When signing up to the tool, participants would complete a short demographic questionnaire that would be used during data capture to link ads to demographic characteristics. Users are also assigned a unique key that links their particular experience of Facebook to a dedicated marker. This key allows them to use the project website to see their individual ad streams (which are available uniquely to them). Once the tool is installed, linked and activated, the tool would allow volunteers to automatically forward ads to us that Facebook sends to them. For each ad, the collected data includes the associated image, image 'carousels' (rotating sets of ad images) or, in the case of advertisements with videos, the first frame of the video. We also collect the name of the account that produced the ad, as well as the associated copy-written portion of the ad and the URLs for the links incorporated into the ad. The ads are displayed in the database in the order they appear, with the most recent at the top of the list of collected ads.

The tool allows us to tag ads manually by category which can then serve as the basis for automated filtering, so that we can, for example, search for all the ads in the 'technology' category to see how these are targeted by age, gender, geography, education level, and so on. For example, in a pilot study with 150 participants, we found a strong demographic skew toward men in the delivery of technology ads: 2,519 views compared to 741 for women. This does not give us information about how many unique users viewed the ads, but the tool can be adjusted to count either unique viewers or the overall number of ad impressions. This functionality is useful for demonstrating how, for example, existing cultural biases are incorporated into advertising patterns. The content tags are editable, and new tags can be added as different categories of ads appear. Ads can be tagged in multiple categories where appropriate. As data collection scales up, it will be increasingly difficult to tag ads by hand, which means that it will become necessary to rely on automated classification systems that categorise content based on text and images.

Automated analysis of automated ads

One of the defining challenges of providing accountability for online advertising is the ability to discern targeting patterns at scale. Companies such as Facebook and Google have this data, but they are not required to share it publicly. Until this happens – and it should – we are left with only rare glimpses of how these giant corporations are shaping our media environment. Because of the possibility for illegal discrimination and the threat to public and democratic culture, these companies should be subject to advertising audits by independent organisations that have full access to their data on ad delivery and targeting. Any attempt to 'reverse engineer' targeting at scale confronts the issues raised by the sheer number of ads that flood online platforms. In the case of our pilot project, even a relatively short ad collection period for a small group yielded more than 10,000 ads – a number that is already unwieldy for analysing without some form of automated information processing. We anticipate then that tools such as ours would require the use of dedicated image and text classification systems.

Such systems could also be used to provide insights into how images and language, in addition to the ads themselves, are demographically targeted. For example, we were able to partner with collaborators who have developed an automated image classification system to explore how the composition of ad images aligned with demographic categories. The first step in the analysis was to create a database that included all of the images collected by the ad collector. We then used the ad classifier to identify clusters of images that were similar to one another but not identical.

In our pilot study, the classifier identified 66 clusters in our data set, although the size of these varied from just a few images to scores of them. We left in repeated images in order to get a sense of the overall volume of ads served and to address the fact that the same ad may have been served to different users. Once the tool created the clusters, we could get a demographic breakdown of each cluster based on the information provided by our participants. This breakdown allowed us to visually detect the demographic skew of ad clusters. The descriptions of the particular clusters, such as 'sleek car' – which comprised ads that featured close up of parts of cars – glossy windshields and shiny grilles – were done manually. We inspected the cluster to get a sense of what elements the photos it contained had in common with one another. We could see from each cluster, thanks to the data visualisation, which demographic characteristics were associated with it.

At this point, the analysis takes on a qualitative component. The image classifier allows us to ask how a particular image cluster might line up with the associations attached to particular demographic profiles. We note that this is a matter of interpretation, but it is an important aspect of assessing how associations and stereotypes might be reproduced by targeted advertising. For example, the 'sleek' looking car ads cluster has a decidedly masculine skew in terms of those participants who were targeted by this cluster. By contrast, the 'Sleeping Related' cluster composed of ads featuring images of domestic life had a very strong female skew.

It is not difficult to see some of the coded differences between these two sets of ads. The car ads focused on abstracted car parts, featured dark colours, fetishising technology, and do not include images of people or sociality. By contrast, the sleeping cluster features softer colours, interiors, images of family life and domesticity. We might compare this with the 'Dining' cluster, which also features scenes of domestic sociality and was seen primarily by participants who identified as female.

It is possible to access all of the images to gain a sense of the overall character of the cluster. It is also possible to isolate particular images and see the demographic characteristics of those who received them. This sample of images provides an example of how the data collected by the ad collection tool might reveal the ways in which design elements and their social associations are distributed demographically. We focus on automated image analysis because of the role it might play in raising awareness about ad targeting and in advancing the discussion about forms of stereotyping reinforced by commercial messaging strategies. A range of tools can be enlisted for making sense of the data collected by the tool. For example, text-based analysis might indicate which terms are most likely to be used in advertising to different groups. Such forms of automated data analysis would rely on the ability to collect large amounts of data from a wide array of participants.

The tool also includes a potentially useful element for promoting media literacy and enabling qualitative research. Participants who install the Facebook ad extension can access our database, enter their personal code and view the history of ads they received while browsing Facebook. These ads, as we have discovered in our follow-up conversations with participants, provide useful discussion points for considering how Facebook interprets the actions and interests of users. This capability can also be used to prompt discussions about which actions Facebook noticed and recorded, leading users to speculate about why they might have received particular ads, and, more pointedly, highlighting just how detailed a portrait of their personal lives Facebook is assembling. One of our participants, for example, speculated upon looking through her ad stream, that she was being targeted with a series of ads for special education services because one of her children is on the autism spectrum – although this is not something she had posted about or discussed on Facebook. She had, however, searched for information online about autism, and Facebook is able to collect information about online activity that takes place beyond the platform.

The abstract knowledge that Facebook was targeting ads based upon users' online behaviour became concrete for our respondents, who saw which interests were prioritised by advertisers, including, for example, gambling and travel – two very active ad categories (for research conducted prior to the COVID-19 pandemic restrictions). These interests did not necessarily coincide with the types of services and products respondents said they were most interested in – but marked the convergence of inferences made from their online explorations with the agenda of advertisers. The most frequently seen ad categories included health and fitness, real estate, education and finance. Despite the promise of the power of targeting based on social network data, respondents were often mystified by the ads they received, noting that they did not coincide with either expressed or latent interests. This response highlighted the false expectation set by the industry: that somehow all ads would become pressingly relevant thanks to the power of the database. It also highlighted the fact that many ads continue to pass unnoticed – along with the patterns they form. Even in the relatively short period of time covered by the pilot study (six weeks), the ads did trace shifts over time. What emerged was not a comprehensive portrait but what might be described as a discontinuous narrative that fixated on life events, passing interests and emerging concerns to form a dynamic, slightly random, portrait of individuals throughout the course of their increasingly mediated lives.

Conclusion

In the world of online advertising, we run the risk of no longer experiencing mediated representations – even those that take the forms of advertising – as social phenomena but as matters of individual taste and personal preference. Even as advertisers promote the promise of customisation and individualisation to consumers, however, they offer a different one to their clients: the ability to provide a top–down form of messaging and influence. The claim of increased advertising effectiveness (via detailed consumer tracking and ad testing) is coupled with that of opacity: the ability to fly under the radar of public scrutiny. Our pilot project was too small in scope to justify making overarching claims about the ways in which online advertising challenges or resuscitates existing stereotypes and associations, although the initial results provide some insight into how targeting reinforces existing gender stereotypes regarding both the content and composition of ad messaging. Much more extensive data would be needed to get a sense of the robustness of these patterns and to discern more nuanced ones based on a range of demographic categories. We hope to be able to recruit participants on a larger scale and to form collaborations with schools, universities, civil rights organisations, labour unions and others to gain more visibility into the contemporary advertising ecosystem. The function of tools like the ones we have described here is not simply to hold advertisers accountable – although this is crucially important – but also to work toward the process of re-socialising the increasingly fragmented realm of cultural representation. We also hope that tools like this and the collaborations they enable might contribute to the ongoing conversation regarding how best to regulate online advertising, which lies at the heart of the contemporary information economy.

Acknowledgments

This work was supported by the Australian Communications Consumer Action Network (grant # 20192009), and by the Australian Research Council (grant # CE200100005).

References

Ali M, Sapiezynski P, Bogen M, Korolova A, Mislove A and Rieke A (2019) Discrimination Through Optimization: How Facebook's Ad Delivery Can Lead to Biased Outcomes. *Proceedings of the ACM on Human-Computer Interaction* 3: 1–30.

Angwin J, Tobin A and Varner M. (2017) Facebook (Still) Letting Housing Advertisers Exclude Users by Race. *ProPublica*, 21 November. Available at: www.propublica.org/article/facebook-advertising-discrimination-housing-race-sex-national-origin (accessed 10 July 2021).

Appen (2021) How to Reduce Bias in AI with a Focus on Training Data. Available at: https://appen.com/blog/how-to-reduce-bias-in-ai/ (accessed 14 May 2021).

Billig M (1995) *Banal Nationalism*. London: SAGE.

Booker B (2020) Uncle Ben's Changing Name to Ben's Original After Criticism of Racial Stereotyping. *National Public Radio*, 23 September. Available at: www.npr.org/sections/live-updates-protests-for-racial-justice/2020/09/23/916012582/uncle-bens-changing-name-to-ben-s-original-after-criticism-of-racial-stereotypin (accessed 2 July 2021).

Carr S (2021) How Many Ads Do We See a Day in 2021. *PPC Protect*, 15 February. Available at: https://ppcprotect.com/blog/strategy/how-many-ads-do-we-see-a-day/ (accessed 10 July 2021).

Ewen S (1974) *The Captains of Consciousness: The Emergence of Mass Advertising and Mass Consumption*. Albany, NY: State University of New York at Albany Press.

Goebel R, Chander A, Holzinger K, Lecue F, Akata Z, StumPf S, Kieseberg P and Holzinger A (2018) Explainable AI: The New 42? Conference Paper in *International Cross-Domain Conference for Machine Learning and Knowledge Extraction*. Berlin: Springer, 295–303. Available at: https://link.springer.com/chapter/10.1007/978-3-319-99740-7_21 (accessed 10 July 2021).

Green K and Issenberg S (2016) Inside the Trump Bunker, With Days to Go. *Bloomberg Business*, 27 October. Available at: www.bloomberg.com/news/articles/2016-10-27/inside-the-trump-bunker-with-12-days-to-go

Jackson K and Barnes J (2015) Representation and Power: A Picture Is Worth a Thousand Words – "Nulla-Nulla: Australia's White Hope, The Best Household Soap' 1920s. *Journal of Australian Indigenous Issues* 18(1): 70–4.

Kaya Yurieff C (2021) Facebook Settles Lawsuits Alleging Discriminatory Ads. *CNN*. Available at: https://edition.cnn.com/2019/03/19/tech/facebook-discriminatory-ads-settlement/index.html (accessed 14 May 2021).

Kilbourne J (1990) Beauty and the Beast in Advertising. In: Scott B, Cayleff S, Donadey A and Lara I (eds) *Women in Culture: An Intersectional Anthology for Gender and Women's Studies*. Baltimore, MD: John Wiley & Sons, 183–86.

Larson J, Angwin J and Valentino-DeVreis J (2017) How We are Monitoring Political Ads on Facebook. *ProPublica*, 5 December. Available at: www.propublica.org/article/how-we-are-monitoring-political-ads-on-facebook (accessed 5 May 2021).

Larson S (2017) Discriminatory Ads Still Get through on Facebook, Investigation Finds. *CNNMoney*. Available at: https://money.cnn.com/2017/11/21/technology/facebook-ads-discriminatory-renting-propublica/index.html?iid=EL (accessed 14 May 2021).

Lears J (1995) *Fables of Abundance: A Cultural History of Advertising in America*. New York: Basic Books.

Marchand R (1985) *Advertising the American Dream: Making Way for Modernity*, 1920–1940. Vol. 53. Stanford, CA: University of California Press.

Sandvig C, Hamilton K, Karahalios K and Langbort C (2014) Auditing Algorithms: Research Methods for Detecting Discrimination on Internet Platforms. Paper presented at: Data and Discrimination: Converting Critical Concerns into Productive Inquiry 22. Available at: http://social.cs.uiuc.edu/papers/pdfs/ICA2014-Sandvig.pdf (accessed 1 May 2021).

Saunders J (2020) Dark Advertising and the Democratic Process. In: Macnish MK (ed) *Big Data and Democracy*. Edinburgh: Edinburgh University Press, 73–85.

Schudson M (1984) *Advertising: The Uneasy Persuasion*. New York: Basic Books.

Statista (2021) Number of Monthly Active Facebook Users Worldwide as of 2nd Quarter 2021. Available at: www.statista.com/statistics/264810/number-of-monthly-active-facebook-users-worldwide/ (accessed 10 October 2021).

The Media Business (1989) *The New York Times*, 1 May. Available at: www.nytimes. com/1989/05/01/business/the-media-business-an-uproar-over-billboards-in-poor-areas.html (accessed 1 May 2021).

Wilson C II and Guttierez F (1995) *Race, Multiculturalism, and the Media: From Mass to Class Communication.* London: SAGE.

INDEX

Note: italic page numbers indicate figures.